ON CUTTING OFF A RATIO

Apollonius of Perga

ON CUTTING OFF A RATIO

An Attempt to Recover
the Original Argumentation
through a
Critical Translation
of the
Two Extant Medieval Arabic Manuscripts

Translated by
E. M. Macierowski

Edited by
Robert H. Schmidt

The Golden Hind Press

Published by:
The Golden Hind Press
Fairfield, Connecticut

EDITOR'S INTRODUCTION

E. M. Macierowski began this translation of the two Arabic manuscripts at the request of The Golden Hind Press. Shortly thereafter he expanded the scope of the project to include a critical edition of these texts. He applied to NEH for funding of his work and was awarded grant RL-20907-86.

The first draft of this translation was a literal one, and posed us with a problem. This work is the only surviving Greek treatise that can show how diorismos was an integral part of the Greek art of analysis & synthesis. It had been our hope that it would challenge mathematicians with its sophisticated geometrical thinking on a rather neglected topic. But even though the logical moves of the arguments, and the very grammatical structure of some of the sentences, seemed to hint at a coherent method, the words of the Arabic translator were often discordant with the argumentation itself.

This is why we conceived the plan of using the literal translation as a basis for reaching back to the argumentation of the lost Greek original, a further "critical translation" if you will. Macierowski himself wanted to take part in this experiment, and agreed to collaborate with us on this task.

Macierowski is continuing his work on the critical edition and literal translation, which we are eagerly awaiting, as it promises to expose the new conceptual beginning that analysis & synthesis had in its Arabic setting.

Justification for the analytical steps in Apollonius' arguments will be found in the propositions of Euclid's *Recipients, Commonly Called The Data*, a primary work also published by The Golden Hind Press. *Recipients* is accompanied by a detailed introduction which attempts to elucidate fully the method of analysis and synthesis as practiced by the Greek geometers. The same points are also made in summary form at the end of this book. The italicized editorial commentary woven into the first locus develops some of these points in context.

The translator's remarks on the methods used in this translation, the use made of the manuscripts, and the departures from a literal translation, will be found after the translation itself. He has also compiled a three column glossary of the most important terms, correlating the translated Arabic word with its presumed Greek antecedent.

Square brackets, [ ], in the translation indicate restored text. Pointed brackets, < >, indicate that a significant editorial decision was made.

ON CUTTING OFF A RATIO

Two unbounded straight lines being positioned in a plane, either parallel or intersecting, on each of which a point is recipient, and a ratio being given, and a point that is not on either of the lines being recipient: to draw a straight line from the recipient point across the positioned lines, cutting them off so that the ratio to one another of the two segments adjacent to the two recipient points on the two lines is the same as the given ratio.

The term 'recipient' here corresponds to the Greek participle commonly translated as 'given', which is indeed a translation appropriate to its syntax in an ordinary Euclidean problem. However its apparent syntax in the above problem resembles its syntax in analytical argumentation, which requires a different translation. The above problem is motivated by an analysis and has to be done in the context of analysis. It need not be thought of as a problem that has to be subjected to analysis. Analysis & synthesis was not merely–or even originally, I maintain–a problem solving art. Here the analytic mode is adopted for the sake of a diorismic inquiry, meaning that it seeks to know what limitations apply to the ratios cut off by the line in its various cases. Such an inquiry obviously has importance for determining when, how, and in how many ways the corresponding Euclidean problem is possible, but this does not preclude its having an interest of its own. The above problem has to be done in order to verify that the diorismic work has been correctly carried out. The individual syntheses for each case serve this end.

BOOK ONE

First let the two lines be parallel, and let them be lines AB, CD. Let the recipient point on line AB be point E, and on CD point G. And let the line which passes through these terminal points on the lines be EH. Then the recipient point is either within angle DGH, or within angle BEG & angle DGE, or within angles in adjacent positions.

Locus One. And first let it be within the angle DGH as point F. Then the lines that are drawn from point F, which cut off from the two lines adjacent to the two points E & G segments whose ratio to one another is the same as a given ratio, obtain in three ways. Either they cut off segments from EB, GD; or from EA, GD; or from EA, GC.

Case 1. So let the line be drawn and fall in the first way as line FK, cutting off from the two lines EB, GD a ratio of EK to GL the same as a given ratio. *It is clear that, among all the ratios which are cut off by lines drawn in accordance with this case, there are infinitely many which are the same as ratios that we can produce ourselves by determinate constructions performed on the points & lines originally available to us; this is what it means to be 'given'. The only assumption we make is that the line FK is some one of these. We do not assume, as is usually done, that the problem has been hypothetically done; that is, that the line has somehow been drawn so as to accomodate some particular preassigned ratio.* And let the line EF be joined. So EF has been positioned, *meaning that it has entered into such a state, due to the conditions under which it was introduced, that it would keep to the same place if it were drawn again in accordance with these same conditions.* But CD also is positioned, *and so is already in the aforementioned state.* So point M is recipient, *which is another way of saying that it has been positioned; namely, that it has been provided for with respect to its position.* But each of the two points E,F is

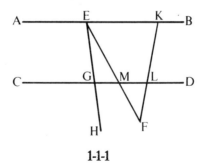

1-1-1

recipient. So the ratio of EM to MF is recipient, *since it has been provided for with respect to its greatness, which will not change as long as the points remain positioned.* And by composition, the ratio of EF to FM is recipient. And the ratio of EF to FM is like the ratio of EK to ML. So the ratio of EK to ML is recipient. But the ratio of EK to GL is recipient. So the ratio of GL to ML is recipient. And by separation, the ratio of GM to ML is recipient. And the line GM is recipient in magnitude, since points G & M are both recipient. *That is, drawing the line through two positioned points always provides it with the same magnitude.* So the line ML is recipient in magnitude and position. But point M is recipient. So L too is recipient. But point F also. So line FK is recipient.

So the line drawn cutting off the ratio occupies the same place as the line drawn through points F & L, which is a line that we can construct out of our own resources. The argument has detached–that is, analyzed–the place of the line from the line itself. The place of the line is no longer dependent upon the original supposition under which the line was introduced.

And since ML is less than GL, the ratio of EK to ML is greater than the ratio of EK to GL. But the ratio of EK to ML is like the ratio of EF to FM. So the ratio of EF to FM is greater than the ratio of EK to GL, which is <the same as> a given ratio. So for that reason it is necessary that the ratio being given <for

the synthesis>–*we are still talking about the given ratio of above*–be less than the ratio of EF to FM. *And thus we detach a limit for the ratio.*

It is only now that we introduce the ratio, namely, the given ratio that was the same as the ratio cut off in the analysis. When a ratio has been provided out of the lines and points originally available to us–that is, when it has been given–it is done with some purpose in mind. In an ordinary geometrical problem the ratio would be offered for the construction of some figure. Here it is offered for synthesis, for restoring the place of the line to the line itself. This is done by drawing the line that was detached in the analysis, showing that it cuts off the given ratio, and then showing that it is the only line that does cut off this ratio. If these conditions are fulfilled, then this line coincides with the one first drawn in the analysis, and the sought line has been put together with its place.

And the synthesis of this problem is as follows: With the same figure, let EF be joined. It has become clear that the ratio being given must be less than the ratio of EF to FM. So let it be the ratio of N to JO, and let it be made that, as the ratio of EF to FM, so the ratio of N to JP, and as the ratio of OP to PJ, so the ratio of GM to ML. And let FL be joined, and produced in a straight line. I say that the line FK alone does what the problem requires. This will be shown as follows.

Since the ratio of GM to ML is like the ratio of OP to PJ, by composition, the ratio of GL to LM is like the ratio of JO to JP. Inversely, the ratio of LM to GL is like the ratio PJ to JO. But also, since the ratio of EF to FM is like the ratio of EK to ML, the ratio of EK to ML is also like the ratio of N to JP. But the ratio of ML to LG is like the ratio of PJ to JO. So, by way of equality, the ratio of EK to GL is like the ratio of N to JO. Thus, from point F a straight line FK has been drawn that cuts off a ratio EK to GL the same as the given ratio. Thus line FK does what the problem requires.

And I say that it alone does this. For, if it be possible for another to do it, then let line FX be drawn cutting off a ratio EX to GR the same as the given ratio. So, since LM is less than

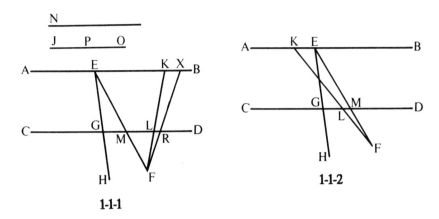

1-1-1

1-1-2

LG, the ratio of RL to LM is greater than the ratio of RL to LG. And by composition, the ratio of RM to ML is greater than the ratio of RG to LG. And the ratio of RM to ML is like the ratio of XE to EK. So it is greater than the ratio of RG to LG. And by alternation, the ratio of XE to RG is greater than the ratio of EK to GL. So line FX does not do what the problem requires. Likewise it is clear that no line other than FK does. Therefore, FK alone does what the problem requires. And it is clear that the lines nearer point G cut off ratios less than those farther from it.

Case 2. With the same things supposed, let line FK be drawn in the second way, cutting off from the two lines EA, GD a ratio of EK to GL the same as a given ratio. And let line EF be joined. So EF has been positioned. But CD is positioned. So point M is positioned. And so each of EF, FM is recipient. So the ratio of EF to FM is a recipient. But the ratio of EF to FM is like the ratio of EK to LM. So the ratio of EK to LM is a recipient. But the ratio of KE to GL is a recipient. So the ratio of GL to LM is a recipient. And by composition, the ratio of GM to ML is a recipient. But the line GM is recipient in magnitude. And point M is recipient. But also point L. So line KF has been positioned. And because line EK can be either equal to LG, or greater than it, or less than it, the ratio is not limited.

And this problem will be synthesized as follows: Let the same things be supposed, and let line EF be joined. And let the given ratio be the ratio of N to JO. And let the ratio of N to JP be made like the ratio of EF to FM. And let it be made that, as the ratio of OP to JP, so the ratio of GM to ML. And let FL be joined and produced in a straight line to K. I say that the line FK does what the problem requires. That is, that the ratio of KE to GL is like the ratio N to JO.

For, the ratio of EF to FM is like the ratio KE to ML. And it is also like the ratio of N to PJ. So the ratio of KE to LM is like the ratio N to PJ. And also, because the ratio of OP to PJ is like the ratio of GM to ML, so, by separation and inversion, the ratio of LM to LG is like the ratio of PJ to JO. And so, by way of equality, the ratio of KE to GL is like the ratio of N to JO. So the line FK does what the problem requires.

And I say that it alone does so. For, if possible, let a line other than it be drawn, and let it be the line FX, which cuts off a ratio of XE to RG the same as the given ratio. Because XE to GR is greater than the ratio of EK to GL, the line XF cuts off a ratio greater than the line KF does. And it is clear that the lines near to point L cut off ratios greater than the ratios that distant lines cut off.

Case 3. Again let everything be the same, only let there be drawn, in the third way, a line FK, cutting off from the two lines EA, GC a ratio of EK to GL the same as a given ratio. Then let EF be joined. So EF is a recipient in position. But CD is a recipient. And so point M is a recipient. And each member of the ratio EF to FM is a recipient. And the ratio of EF to FM is like the ratio of EK to LM. So the ratio of [EK] to LM is a recipient. But the ratio of EK to GL is a recipient. So the ratio of ML to GL is a recipient. And by separation, the ratio of MG to GL is a recipient. But MG is a recipient. So GL is recipient in magnitude. But point G is positioned. And since GL is less than LM, the ratio of EK to GL is greater than the ratio of EK to LM. But the ratio of EK to LM is like the ratio of EF to FM. And so the ratio being given in the synthesis must be greater than the ratio of EF to FM.

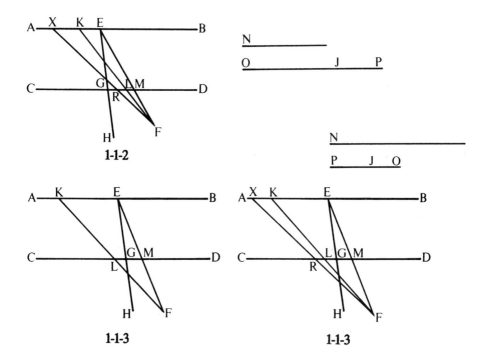

1-1-2

1-1-3

1-1-3

And this problem will be synthesized thus: With things remaining the same, let EF be joined. It is required that the ratio being given be greater than EF to FM. So let the ratio of N to JP be greater than the ratio of EF to FM. And let it be made that, as the ratio of EF to FM, so the ratio of N to OP, and, as the ratio of OJ to JP, so the ratio of MG to GL. Let FL be joined, and let it be produced in a straight line. I say that the line FK does what the problem requires.

For, since the ratio of MG to GL is like the ratio of OJ to JP, so, by composition, the ratio of ML to GL is like the ratio of OP to JP. And also, since the ratio of EF to FM is like the ratio of EK to ML and also like the ratio of N to OP, while the ratio of ML to LG is like the ratio of PO to JP, then, by way of equality, the ratio of EK to GL is like the ratio of N to JP.

Then I say that line FK alone does what the problem requires. For if it be possible for another line to do so, then let another line FX be drawn cutting off a ratio of XE to RG the same as the given ratio. Since line LG is less than line LM, the ratio of RL to LG is greater than the ratio of RL to LM. And by composition, the ratio of RG to GL is greater than the ratio of RM to ML. And the ratio of RM to ML, like the ratio of XE to EK, is less than the ratio of RG to GL. And by alternation, the ratio of XE to RG is less than the ratio of KE to LG. So line FK cuts off a ratio greater than the ratio that FX cuts off. So the lines near point G cut off ratios greater than the ratios cut off by the lines distant from it.

The individual analyses have determined boundaries for all the ratios that can possibly be cut off by the lines in the various cases. It must now be determined whether any arbitrarily chosen ratio within one of these domains is cut off by a line in the corresponding case, thus verifying that the proposed limit is the true limit of the ratios; that is, that there are no undiscovered gaps in the possible ratios cut off. If there were such gaps, we would not have the true limitation of the ratio, and further sublimits would have to be imposed. In effect, just as the syntheses of the individual cases put the sought line back together with its place, here the proposed limit of the ratios must be synthesized, or reimposed on the ratios actually cut off within a case.

So we can now deal with this problem in all its <cases>. But let us clarify, in how many ways the synthesis occurs in the problem for each of its <domains>. Let things be the same as they were, and let line EF be drawn and produced in a straight line to M. The given ratio is either less than the ratio of EF to FM, or the same as it, or greater than it. If the given ratio is less than the ratio of EF to FM, the problem is synthesized in two <ways>, namely according to the first and the second cases. It is not possible for it to be synthesized according to the third case, because the given ratio is not greater than the ratio of EF to FM. And if it is the same as the ratio of EF to FM, then the problem will be synthesized according to the second case. It is not possible for it to be synthesized according to the first case, because it is not greater than the ratio of EF to FM; nor according to the third, because it is not greater than the ratio of EF to FM. And if it is greater than the ratio of EF to FM,

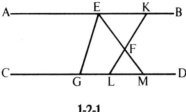

1-2-1

then the problem will be synthesized in two ways, namely, according to the second and third cases. It is not possible for it to be synthesized according to the first, because it is not less than the ratio of EF to FM.

Locus Two. And now let the recipient point be within the two angles BEG & EGD as point F. The lines that are drawn from point F, which cut off lines in a ratio like a given ratio adjacent to the two points E & G, obtain in three ways. Either they cut off from the two lines EB, GD; or from the two lines EA, GD; or from the two lines EB, GC.

Case 1. So let the line fall in the first way, and let a line KL be drawn cutting off from the lines EB, GD the ratio of EK to GL, which is the same as a given ratio. Let EF be joined and produced in a straight line to M. So the line EM is positioned. But the line CD is positioned. So the point M is recipient. And so the ratio of EF to FM is a recipient. And the ratio of EF to FM is like the ratio of EK to ML. But the ratio of EK to LG is a recipient. So, by way of equality, the ratio of GL to ML is a recipient. So, by composition, the ratio of GM to ML is a recipient. But the line GM is a recipient. So the line ML is a recipient in magnitude. But point M is positioned. So point L is positioned. And because the two lines GL and LM can be either equal, or one greater than the other, the ratio has no limitations.

And this problem will be synthesized thus: With things remaining the same, let EF be joined and produced in a straight line to point M. And the given ratio is the ratio of N to JO. Let it be made that, as the ratio of EF to FM, so N to PJ, and, as the ratio of PJ to OJ, so the ratio of ML to GL. And let LF be joined and produced in a straight line. I say that the line LK does what the problem requires.

For, since the ratio of EF to FM, that is, the ratio of EK to LM, is like the ratio of N to PJ, and the ratio of LM to LG is like the ratio of PJ to JO, since it is so by construction: therefore, by way of equality, the ratio of EK to GL is like the ratio of N to JO. And so line KL does what the problem requires.

Now I say that it alone does so. For, if it be possible for another to do so, then let another line such as XF be drawn. And since line EK is greater than line XE, and line GL is less than line RG, the ratio of EK to GL is greater than the ratio of EX to RG. So the line LK cuts off a ratio greater than the ratio that line XR cuts off. For that reason, the lines standing nearer point G cut off ratios that are greater than the ratios that the lines more distant cut off.

Case 2. And let there also be drawn, in the second way, a line KL that cuts off from the two lines EA, GD a ratio of EK to GL the same as a given ratio. Let EF be joined and produced in a straight line. So line EM is positioned. But CD is positioned. So point M is positioned. And so the ratio of EF to FM is a recipient. And the ratio of EF to FM is like the ratio of KE to ML. But the ratio of EK to GL is a recipient. So the ratio of GL to ML is a recipient. So, by separation, the ratio of GM to ML is a recipient. But the line GM is a recipient. So line ML is recipient in magnitude. But point M is a recipient. So point L is a recipient. But point F is positioned. So line KF is positioned. So, because line GL is greater than line LM, the ratio of EK to LM, that is, EF to FM, is greater than the ratio of EK to GL. And the ratio of EK to GL is the same as the given ratio. So the ratio being given in the synthesis must be less than the ratio of EF to FM.

The problem will be synthesized thus: With the same things supposed, and the given ratio the ratio of N to JO, which is less

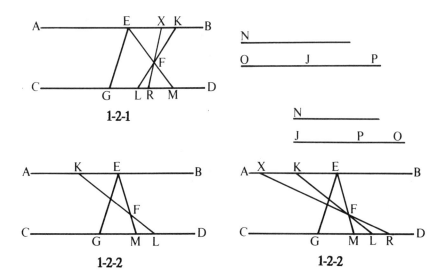

1-2-1

1-2-2

1-2-2

than the ratio of EF to FM, let the ratio of N to PO be made
like the ratio of EF to FM, and let it be made that, as the ratio
of JO to PO, so the ratio of GL to ML. And let LF be joined
and produced in a straight line. Then I say that line LK does
what the problem requires.

Since the ratio of EF to FM, that is, the ratio of EK to ML,
is like the ratio of N to PO, and the ratio of ML to LG is like
the ratio of PO to OJ, so, by way of equality, the ratio of EK to
LG is like the ratio of N to JO. So line KL does what the
problem requires.

And I say that it alone does so. For if it be possible for
another to do so, then let another line be drawn, such as XR.
Now, since line LG is greater than LM, the ratio of RL to LG
is less than its ratio to line LM. And by composition, the ratio
of GR to GL is less than the ratio of RM to ML. But the ratio
of RM to ML is like the ratio of XE to EK. So the ratio of GR
to GL is less than the ratio of XE to EK. But by alternation,
the ratio of GR to XE is less than the ratio of GL to EK. So
the line KL alone does what the problem requires. And it is
clear that the lines near point G cut off ratios greater than the
lines distant from it.

Case 3. And let a line KL be drawn in the third way, cutting off from the two lines EB, GC a ratio of EK to GL the same as a given ratio. Let EF be joined and produced in a straight line. So EM is positioned. And MD also is positioned. So point M is a recipient. But the two points E, F are recipients. So the ratio of EF to FM is a recipient. And the ratio of FE to FM is like the ratio of EK to LM. So the ratio of EK to LM is a recipient. But the ratio of EK to GL is a recipient. So the ratio of ML to LG is a recipient. And by separation, the ratio of MG to LG is a recipient. But line MG is a recipient. So line GL also is a recipient in magnitude. But point G is a recipient. So point L is a recipient. But F also is a recipient. So line KL is positioned. And because line LG is less than the line LM, the ratio of EK to GL is greater than the ratio of EK to LM. But the ratio of EK to LM is like the ratio of EF to FM. So the ratio of EK to GL is greater than the ratio of EF to FM. But the ratio of EK to GL is the same as the given ratio. So the ratio being given in the synthesis must be greater than the ratio of EF to FM.

And this problem will be synthesized thus: With the figure remaining the same, let the given ratio be the ratio of N to JO, greater than the ratio of EF to FM. And let it be made that, as the ratio of EF to FM, so the ratio of N to PO, and, as the ratio of PJ to JO, so the ratio of MG to GL. And let LF be joined and produced in a straight line. Then I say that the line LK does what the problem requires.

For, since the ratio of MG to GL is like the ratio of PJ to JO, then, by composition, the ratio of ML to GL is like the ratio of PO to JO. And also, since the ratio of EF to FM, that is, the ratio of EK to LM, is like the ratio of N to PO, and the ratio of ML to GL is like the ratio of PO to OJ, so, by way of equality, the ratio of EK to LG is like the ratio of N to JO. So the line KL does what the problem requires.

And I say that it alone does so. For, if it be possible for another line to do so, then let another line XR be drawn. Now, because line ML is greater than GL, the ratio of RL to LM is less than the ratio of RL to LG. So, by composition, the ratio of RM to ML is less than the ratio of RG to GL. But the ratio of RM to ML is like the ratio of XE to EK. So the ratio of XE

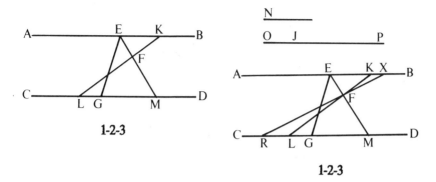

1-2-3

1-2-3

to EK is less than the ratio of RG to GL. And by alternation, the ratio of XE to RG is less than the ratio of EK to GL. So the line LK alone does what the problem requires. And it is clear that the lines next to point G cut off ratios greater than the ratios that the lines distant from it cut off.

Since we have found the <status> of the problem in each of its three cases, it must be made clear how the synthesis comes about in each of its domains. Then let things be the same as before, and let EF be joined and produced in a straight line to M. Now, the given ratio is either like the ratio of EF to FM, or greater than it, or less than it. If the given ratio is like the ratio of EF to FM, then the problem will be synthesized in one sense, namely, according to the first case. It will not be synthesized in the second way, because the ratio is not less than the ratio of EF to FM; and it will not be synthesized in the third way, because it is not greater than the ratio of EF to FM. And if the given ratio is less than it, then the problem will be synthesized in two senses, according to the first case and the second case. It will not be synthesized according to the third case, because the ratio is not greater than the ratio of EF to FM. And if the given ratio is greater than the ratio of EF to FM, then the problem will be synthesized in two senses, according to the first case and the third case. It is not possible for it to be synthesized according to the second case, because the given ratio is not less than the ratio of EF to FM.

Let the two positionally recipient lines AB, CD cut one another at point E. And first let them both have a terminal point at the one point E. The recipient point falls either within the angle DEB, or within any other one of the angles. The rule for all the angles is the rule of the angle DEB.

Locus Three. So let it fall within angle DEB as point G. Now, the lines that are drawn from point G, cutting off lines adjacent to point E whose ratio to one another is like the given, obtain in three ways. Either they cut off from AE, ED; or from CE, EB; or from BE, ED.

Case 1. So let a line GH be drawn in the first way, cutting off from the two lines AE, ED a ratio EF to HE the same as a given ratio. And through point G let a line GK be drawn parallel to line ED. So line GK is positioned. But line AB is positioned. So point K is positioned. And because the ratio of FE to EH is a recipient, the ratio of GK to KH is a recipient. However GK is a recipient. So KH is recipient in magnitude. But point K is a recipient. So point H is a recipient. But point G also is a recipient. So line HG is positioned. And it is clear that the ratio being given is less than the ratio of GK to KE.

And the problem will be synthesized thus: The figure will remain as before, and line GK will be drawn parallel to DE. And let the given ratio, namely, the ratio of L to M, be less than the ratio of GK to KE. Let it be made that, as the ratio of L to M, so the ratio of GK to KH. And let HG be joined. Then I say that line HG does what the problem requires.

For, since the ratio of GK to KH is like the ratio of FE to EH, and also like the ratio of L to M, so the ratio of FE to EH is like the ratio of L to M. So line HG does what the problem requires.

And I say that it alone does so. For, if it be possible for another to do so, then let another line be drawn, such as GN. Since line KH is less than line KN, the ratio of GK to KH is greater than the ratio of GK to KN. But the ratio of GK to KH is like the ratio of FE to EH. And the ratio of GK to KN is

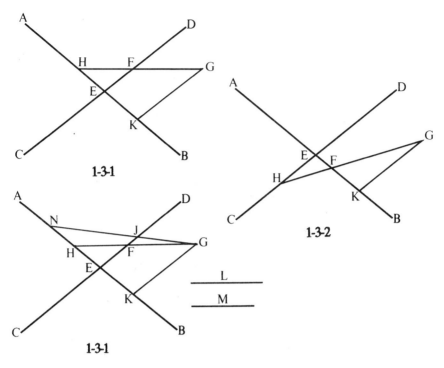

1-3-1

1-3-2

1-3-1

like the ratio of JE to EN. So the ratio of FE to EH is greater than the ratio of JE to EN. So they are not same ratios. And so GN does not separate off a ratio like the given. And likewise it is clear that no other line than GH does so. So line GH alone does what the problem requires. And it is clear that lines near to point E always cut off ratios greater than the ratios that lines distant from it cut off.

Case 2. And let a line GH be drawn in the second way, cutting off from the two lines CE, EB a ratio of HE to EF the same as a given ratio. And let a line GK be drawn parallel to line CD through point G. So GK is positioned. But AB is positioned. So point K is recipient. So the ratio of GK to KF is a recipient. But line GK is a recipient. So KF is recipient in magnitude. But point K is a recipient. So point F also is a recipient. But point G is a recipient. So line GH is positioned. And the ratio being given must be greater than the ratio of GK to KE.

And the problem will be synthesized thus: Things will remain the same, and the given ratio will be greater than the ratio of GK to KE, namely, the ratio of L to M. Let it be made that, as the ratio of L to M, so the ratio of KG to KF, and let GF be joined and produced in a straight line to H. Then I say that GH alone does what the problem requires.

For, since the ratio of GK to KF is like the ratio of HE to EF, and also like the ratio of L to M, so the ratio of HE to EF is like the ratio of L to M. So the line does what the problem requires.

I say that it alone does so. For, if it be possible for another to do so, let another line be drawn, such as GN. Since line KJ is less than line KF, the ratio of GK to KJ is greater than the ratio of GK to KF. However the ratio of GK to KJ is like the ratio of NE to EJ, and the ratio of GK to KF is like the ratio of HE to EF. So the ratio of NE to EJ is greater than the ratio of HE to EF. And so it is not the same as the given ratio. So the line GN does not do what the problem requires. Similarly, it is clear that no other line than GH does what the problem requires. And it is clear that lines near point E always cut off ratios smaller than the ratios that lines distant from it cut off.

Case 3. And let a straight line HF be drawn in the third way, cutting off from the two straight lines BE, ED a ratio of HE to EF the same as a given ratio. And let line GK be drawn parallel to line CD. So line GK is positioned. But line EB is positioned. So point K is a recipient. But the ratio of HE to EF is a recipient. So the ratio of GK to KF is a recipient. But line GK is a recipient. So line KF is recipient in magnitude. But point K is positioned. So point F is positioned. But G is positioned. So line FH is positioned. And the ratio has no limitations, because KE will be either equal to KF, or greater than it, or less than it.

And the problem will be synthesized thus: Everything will be the same with respect to the parallelism. And the given ratio will be the ratio of L to M. Let it be made that, as the ratio of L to M, so the ratio of GK to KF. And let FG be joined and produced in a straight line to H. Then I say that FH alone does what the problem requires.

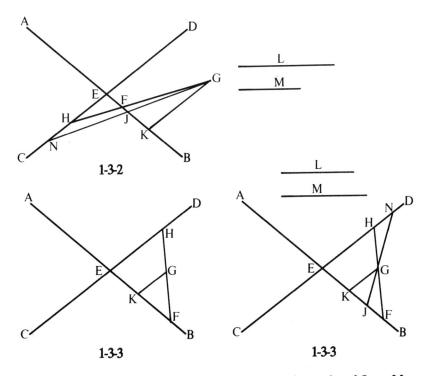

1-3-2

1-3-3   1-3-3

For, since the ratio of GK to KF is like the ratio of L to M, so the ratio of HE to EF is like the ratio of L to M. So line FH does what the problem requires.

And I say that it alone does so. For, if it be possible for another to do so, then let another line be drawn, such as NJ. And since the line NE is greater than the line EH, and line EJ is less than the line EF, so the ratio of NE to EJ is greater than the ratio of HE to EF. And so it is not the same. Likewise it will be clear that no other line than FH will do what the problem requires. And it is clear that the lines near point E, which turn toward CD, take off ratios less than the ratios that lines distant from it cut off.

Since we have explained the <status> of the problem in all its ways, it remains for us to explain in how many ways the problem is synthesized for all its domains. So let things be as before, and let line GK be drawn parallel. Now, the given ratio is either less than the ratio of GK to KE, or greater than it, or the same as it. If the given ratio is less than the ratio of GK to KE, then the problem will be synthesized in two ways, according to the first and third cases. It is not possible for it to be synthesized according to the second case, because it is not greater than the ratio of GK to KE. And if it is greater than the ratio of GK to KE, the problem will be synthesized in two ways, according to the second and third cases. It is not possible for it to be synthesized according to the first case, because it is not less than the ratio of GK to KE. And if it is the same as the ratio of GK to KE, then the problem will obtain in one way, according to the third case. It will not be possible for it to be synthesized according to the first case, because it is not less than the ratio of GK to KE; nor according to the second case, because it is not greater than the ratio of GK to KE.

Let the lines AB, CD cut one another at point E. Let the terminal point on line AB be point G, and on line CD point E. Now, the recipient point is either within angle DEB, or within angle AEC, or within adjacent angles.

Locus Four. Then first let it be within angle DEB as point H. The straight lines that are drawn from point H, taking off lines next to the points E & G in a given ratio, obtain in four ways. Either they will cut off from the two lines ED, GA; or from ED, GE; or from EC, GB; or from ED, GB.

Case 1. So let a line HF be drawn in the first way, cutting off from the two lines ED, GA a ratio of EK to GF the same as a given ratio. And let a line HL be drawn parallel to line DE. So point L is a recipient. And let it be made that, as the ratio of EK to GF, so the ratio of LH to GA. Since the ratio of LH to AG is a recipient, and line LH is a recipient, GA also is in magnitude & position. But point G is a recipient. So point A is a recipient. But point L is a recipient. So line AL is a recipient. And because the ratio of LH to GA is like the ratio of EK to GF, then, by alternation, the ratio of LH to EK is like the ratio of AG to GF. However the ratio of LH to EK is like the ratio

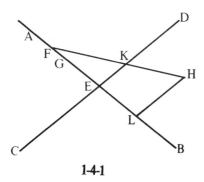

1-4-1

of LF to FE. So the ratio of LF to FE is like the ratio of AG to GF. And by conversion, the ratio of FL to LE is like the ratio of GA to AF. So the <rectangle contained by AG & EL> is equal to <that which is contained by LF & FA>. However <that which is contained by AG & EL> is a recipient. So <that which is contained by LF & FA> is a recipient. And to a recipient line, namely AL, there has been applied a rectangle equal to a recipient rectangle, and deficient by a square. So each of AF, FL is a recipient. So point F is a recipient. But point H is a recipient. So line FH is a recipient.

Now let it be shown that the application is a possibility. It is necessary in the synthesis to apply to the line LA a rectangle equal to the rectangle LE by GA and deficient from the complete line by a square. That is possible since LG by GA is greater than LE by GA.

And this problem will be synthesized thus: Let HL remain parallel as before. And let the given ratio be the ratio of M to N. Let the ratio of LH to GA be made like the ratio of M to N. And let there be applied to line AL a rectangle LF by FA equal to the rectangle AG by EL and deficient by a square. And let FH be joined. Then I say that line FH alone does what the problem requires. That is, that the ratio of M to N is like the ratio of KE to GF.

For, since the rectangle LF by FA is equal to the rectangle

AG by EL, the ratio of FL to LE is like the ratio of GA to AF. And by conversion, the ratio of LF to FE is like the ratio of AG to GF. However the ratio of LF to FE is like the ratio of LH to EK. So the ratio of LH to EK is like the ratio of AG to GF. And by alternation, the ratio of LH to GA is like the ratio of EK to GF. However the ratio of LH to GA was made like M to N. So the ratio of EK to GF is like the ratio of M to N. So line FH does what the problem requires.

I say that it alone does so. For if it be possible for another to do so, let another line be drawn, such as HJ. And because rectangle LG by GA is greater than rectangle LF by FA, rectangle LF by FA is greater than rectangle LJ by JA. And rectangle LF by FA is equal to rectangle GA by LE. So the rectangle GA by LE is greater than the rectangle LJ by JA. So the ratio of JL to LE is less than the ratio of GA to AJ. And the ratio of LJ to JE, by conversion, is greater than the ratio of AG to GJ. However the ratio of LJ to JE is like the ratio of LH to ES. So the ratio of LH to ES is greater than the ratio of AG to GJ. And by alternation, the ratio of LH to GA is greater than the ratio of ES to GJ. However the ratio of LH to GA is like the ratio of M to N. So the ratio of M to N is greater than the ratio of ES to GJ. And it is clear that the lines near to point E cut off ratios greater than the ratios that the lines distant from it cut off.

Case 2. Let a line HF be drawn in the second way, cutting off from the two lines ED, GE a ratio of EK to GF the same as a given ratio. And let line HL be drawn parallel. So line HL is a recipient. So point L is a recipient. And let it be made that, as the ratio of EK to GF, so the ratio of HL to GM. But line HL is a recipient. So line GM is recipient in magnitude. But point G is a recipient. So point M is a recipient. And since the ratio of LH to GM is like the ratio of EK to GF, then, by alternation, the ratio of LH to EK is like the ratio of GM to GF. However the ratio of LH to EK is like the ratio of LF to FE. So the ratio of GM to GF is like the ratio of LF to FE. And by conversion, the ratio of MG to FM is like the ratio of FL to LE. So rectangle GM by EL is equal to rectangle LF by FM. However rectangle GM by EL is a recipient. So rectangle LF by FM is a recipient.

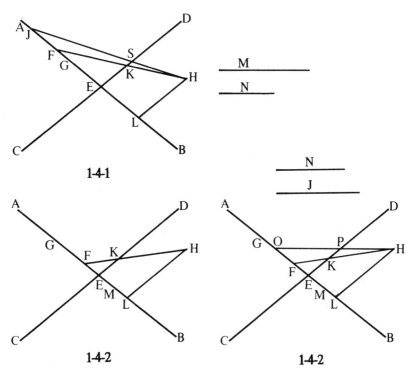

1-4-1

1-4-2

1-4-2

And it has been applied to a recipient line, namely line ML, with the excess of a square. So point F is a recipient. But point H is a recipient. So line HF is positioned.

And this problem will be synthesized thus: Let the parallelism be the same as before. The given ratio is the ratio of N to J. Let the ratio of HL to GM be made like the ratio of N to J. And let there be applied to line ML a rectangle equal to rectangle GM by LE with the excess of a square, namely, rectangle LF by FM. Then, since the rectangle LE by GM has been applied to line LM with the excess of a square, and the rectangle LG by GM is greater than the rectangle LE by GM, so the rectangle equal to rectangle LE by GM, when it is applied with the excess of a square, is not the rectangle LG by GM. So let it be the rectangle LF by FM, and let line FH be drawn. Then I say that line FH does what the problem requires.

And I say that it alone does so. For if it were possible for another to do so, then let another line be drawn, such as HPO. And since line PE is greater than line EK, and line GO is less than line GF, so the ratio of line PE to line GO is greater than the ratio of EK to GF. So line HF [alone] does what the problem requires. And it is clear that lines near to point E cut off ratios less than the ratios that lines distant from it cut off.

Case 3. Let a line KH be drawn according to the third case, cutting off from the lines EC, GB a ratio of KE to FG the same as a given ratio. And through point H let line HL be drawn parallel to CD. So it is clear that the line HL is a recipient in magnitude, and point L is a recipient. And let the ratio of LH to GM be made like the ratio of KE to FG. Since the ratio of KE to FG is a recipient, and it is like the ratio of LH to GM, the ratio of LH to GM is a recipient. But line LH is a recipient. So line GM is recipient in magnitude. But point G is a recipient. So point M is a recipient. But point L is a recipient. So line LM is a recipient. And since the ratio of LH to GM is like the ratio of KE to GF, then, by alternation, the ratio of LH to KE is like the ratio of MG to GF. And the ratio of LH to KE is like the ratio of LF to FE. So the ratio of LF to FE is like the ratio of MG to GF. And by composition, the ratio of LE to EF is like the ratio of MF to FG. And by conversion, the ratio of EL to LF is like the ratio of FM to MG. So rectangle LE by MG is equal to rectangle MF by FL. However rectangle LE by GM is a recipient, because each of the two lines is a recipient. So rectangle MF by FL also is a recipient. And it has been applied to the recipient line ML deficient by a square. So point F is recipient. And it falls between the two points E & L because ME by EL is greater than MG by EL. And point H is recipient. So line HK is positioned.

And the problem will be synthesized thus: Let it be supposed that the parallelism is the same. Let the given ratio be the ratio of N to J, and let there be applied to ML a rectangle equal to rectangle LE by GM and deficient by a square, namely, the rectangle MF by FL. And let line HF be joined and

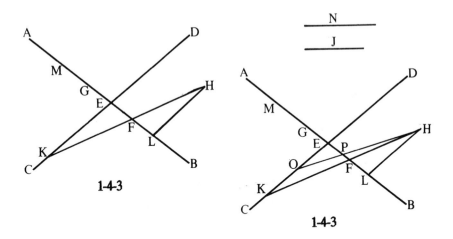

1-4-3

1-4-3

produced in a straight line. Then I say that line HK does what the problem requires. That is, that it cuts off a ratio of KE to GF like the ratio of N to J.

For, since rectangle LE by GM is equal to rectangle MF by FL, the ratio of EL to LF is like the ratio of MF to MG. And by conversion, the ratio of LE to EF is like the ratio of FM to FG. And by separation, the ratio of LF to EF, which is the same as the ratio of LH to KE, is like the ratio of MG to GF. And by alternation, the ratio of LH to MG is like the ratio of KE to GF, and the ratio of LH to GM is like the ratio of N to J. So the ratio of N to J is like the ratio of KE to GF. So line KH does what the problem requires.

Then I say that it alone does so. For if it were not so, let another line be drawn, such as line HO. Then, if line HO cuts off the given ratio, which is the same as the ratio of N to J, the ratio of KE to GF is like the ratio of OE to GP. But the ratio of KE to GF is like the ratio of LH to GM. So the ratio of LH to GM is like the ratio of OE to GP. And by alternation, the ratio of LH to OE is like the ratio of MG to GP, and the ratio of LH to OE is like the ratio of LP to PE. So the ratio of LP to PE is like the ratio of MG to GP. And by composition, the ratio of LE to EP is like the ratio of MP to PG. And by conversion, the ratio of EL to [PL] is like the ratio of PM to MG. So the rectangle LE by GM is equal to the rectangle PM by PL. But the rectangle MF by FL is equal to the rectangle LE by GM. So the rectangle FM by FL is equal to the rectangle MP by PL. This is absurd and impossible. So line KH alone does what the problem requires.

However, it is known [that the rectangle EL by MG] is less than the rectangle MP by PL. So the ratio of EL to PL is less than the ratio of MP to MG. And by conversion, the ratio of LE to EP is greater than the ratio of MP to PG. And by separation, the ratio of LP to PE is [greater than the ratio of MG to GP. And the ratio of LP to PE is] like the ratio of LH to OE. So the ratio of LH to OE is greater than the ratio of OE to GP. But the ratio of LH to GM is like the ratio of KE to GF. So the ratio of KE to GF is greater than the ratio of OE to GP. So line KH cuts off a ratio greater than the ratio that line OH cuts off. So it is clear that lines near to point E cut off ratios less than the ratios that lines distant from it cut off.

Case 4. Let a line KF be drawn in the fourth way, cutting off from the two lines ED, GB a ratio of EF to KG the same as a given ratio. And let line HL be drawn parallel to line CD. So line HL is recipient in position and in magnitude. So point L is recipient. And because the ratio of EF to KG is a recipient, and the ratio of LH to GM is made like it, the ratio of LH to GM is a recipient. But line LH is recipient in magnitude. And point G is recipient. So point M also is recipient. But point L is recipient. So line ML is recipient. And since the ratio of EF to KG is like the ratio of LH to GM, then, by alternation, the ratio of EF to LH, and also the ratio of EK to KL, is like the ratio of KG to GM. And by separation, the ratio of EL to LK is like the ratio of KM to MG. So the rectangle LE by MG is equal to the rectangle MK by KL. However the rectangle LE by MG is a recipient. But it has been applied to a recipient line, namely line ML, with the excess of a square. So point K is recipient. But point H is recipient. So line KF is positioned. And that is what we wanted to show.

The problem will be synthesized thus: With the parallel line remaining the same, let the given ratio be the ratio of N to J,

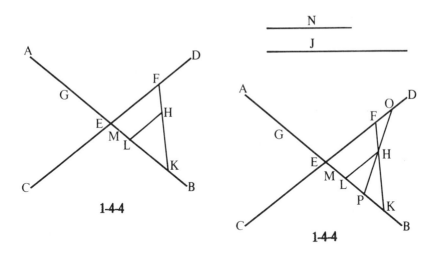

1-4-4

and let the ratio of HL to GM be made the same as the ratio of
N to J. And let there be applied to line ML a rectangle equal to
the rectangle LE by MG and exceeding by a square, namely, the
rectangle MK by KL. And let HK be joined and produced in a
straight line. Then I say that line FK does what the problem
requires. That is, that it cuts off a ratio of EF to GK like the
ratio of N to J.

For, since the rectangle LE by GM is equal to the rectangle
MK by KL, the ratio of EL to LK is like the ratio of KM to
MG. And by composition, the ratio of EK to KL is like the
ratio of KG to GM. But the ratio of EK to KL is like the ratio
of EF to LH. So the ratio of EF to LH is like the ratio of KG
to GM. And by alternation, the ratio of EF to GK is like the
ratio of LH to GM. But the ratio of LH to GM is like the ratio
of N to J. So the ratio of N to J is like the ratio of EF to GK.
So line FG does what the problem requires.

And I say that it alone does so. For if it were possible for
another to do so, let another line, such as OP, be drawn. Then,
if line OP were to cut off a ratio the same as the ratio of N to
J, the ratio of EO to GP would be like the ratio of EF to GK.
However the ratio of EF to GK is like the ratio of LH to GM.
So the ratio of EO to GP is like the ratio of LH to GM. And
by alternation, the ratio of EO to LH is like the ratio of PG to

GM. But the ratio of EO to LH is like the ratio of EP to PL. So the ratio of EP to PL is like the ratio of PG to GM. And by separation, the ratio of EL to LP is like the ratio of MP to MG. So the rectangle LE by MG is equal to the rectangle MP by PL. However the rectangle LE by MG is equal to the rectangle MK by KL. So rectangle MK by KL is equal to rectangle MP by PL. But that is something that is impossible. So line KF alone does what the problem requires.

However, it is known that it cuts off a greater ratio. That will be explained in this way: Since rectangle MK by KL, that is, rectangle LE by MG, is greater than the rectangle MP by PL, the ratio of EL to LP is greater than the ratio of MP to MG. And by composition, the ratio of EP to PL is greater than the ratio of PG to GM. However the ratio of EP to PL is like the ratio of EO to LH. So the ratio of EO to LH is greater than the ratio of PG to GM. And by alternation, the ratio of EO to PG is greater than the ratio of LH to GM. However the ratio of LH to GM is like the ratio of EF to KG. So the ratio of EO to GP is greater than the ratio of EF to KG. And for that reason line KF will cut off a ratio less than the ratio that line OP cuts off. So it is clear that lines near point E cut off ratios less than the ratios that lines farther from it cut off. And that is what we wanted to show.

So the problem will be possible in every case, and it will be synthesized in each way for every domain <of the ratio>, because all the cases are unrestricted.

Let the recipient point be within angle CEA as point H. And through point H let a line be drawn parallel to line AB. The point G falls either above or below or on the parallel line.

Locus Five. Let it first fall on it. The lines drawn from point H obtain in three ways. Either they are drawn so as to cut off a ratio from the lines CG, EA; or from the lines EG, EB; or from the lines EA, GD.

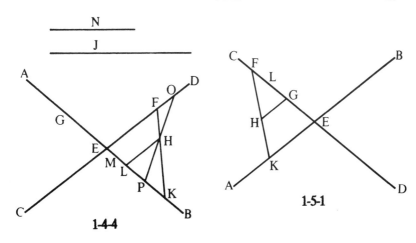

1-4-4

1-5-1

Case 1. So initially let it be drawn in the first way, cutting off from lines EA, CG the ratio of EK to GF the same as a given ratio. And let the ratio of GH to GL be the same as the ratio of EK to GF. So the ratio of HG to GL is a recipient. But line GH is a recipient. So line GL is a recipient. But point G is a recipient. So point L is a recipient. So line GL is a recipient in position. And since the ratio of EK to GF is like the ratio of HG to GL, then, by alternation, the ratio of EK to GH is like the ratio of FG to GL. However the ratio of EK to GH is like the ratio of EF to FG. So the ratio of EF to FG is like the ratio of FG to GL. And by separation, the ratio of EG to FG is like the ratio of FL to LG. So rectangle EG by GL is equal to rectangle GF by FL. And rectangle EG by GL is a recipient, because each of them is recipient. So rectangle GF by FL also is recipient. And to a recipient line, namely LG, it has been applied with the excess of a square. So line GF is recipient. And point H is recipient. So line FK is positioned.

And the problem will be synthesized thus: With the parallel line remaining the same, let the given ratio be the ratio of M to N. And let the ratio of HG to GL be made like the ratio of M to N. And let there be applied to line GL a rectangle equal to the rectangle EG by GL which exceeds by a square, namely, rectangle GF by FL. And let FH be joined and produced in a straight line. Then I say that line FK alone does what the problem requires. That is, that the ratio of M to N is like the ratio of EK to GF.

For, since the rectangle EG by GL is equal to rectangle GF by FL, the ratio of GE to FG is like the ratio of FL to LG. And by composition, the ratio of EF to FG is like the ratio of [FG to LG. And the ratio of EF to FG is like the ratio of] EK to GH. So the ratio of EK to GH is like the ratio of FG to GL. And by alternation, the ratio of EK to FG is like the ratio of HG to GL. However the ratio of HG to GL is like the ratio of M to N. So the ratio of EK to GF is like the ratio of M to N. So the line FK does what the problem requires.

Then I say that it alone does so. For if it were possible for another to do so, let another line be drawn, such as JO. So, if JO cuts off a ratio the same as the ratio of M to N, and the ratio of EK to GO is like the ratio of EJ to GO, and the ratio of EK to GF is like the ratio of HG to GL, then the ratio of GH to GL will be like the ratio of EJ to GO. And by alternation, the ratio of EJ to GH is like the ratio of OG to GL. So the ratio of EJ to GH is like the ratio of EO to GO. So the ratio of EO to OG is like the ratio of OG to GL. And by separation, the ratio of [EG to GO is like the ratio of OL to GL. And so the rectangle] EG by GL is equal to the rectangle GO by OL. And rectangle EG by GL is equal to rectangle GF by FL. So rectangle GF by FL is equal to rectangle GO by OL. But that is not possible. So line FK is the one that does what the problem requires.

However, we know that [it cuts off a lesser ratio. Since line EJ] is greater than line EK, and line FG is greater than line GO, the ratio of EJ to GO is greater than the ratio of EK to GF. So line FK cuts off a ratio less than the ratio that line OJ cuts off. So it is clear that the lines nearer point E cut off ratios less than the ratios that the lines farther from it cut off.

Case 2 Let a line HK be drawn in the second way, cutting off from the lines EB, EC a ratio of KE to FG the same as a given ratio. So the ratio of KE to FG is a recipient. And it is the same as the ratio of HG to GL. So the ratio of HG to GL is a recipient. But line HG is a recipient. But point G is a recipient.

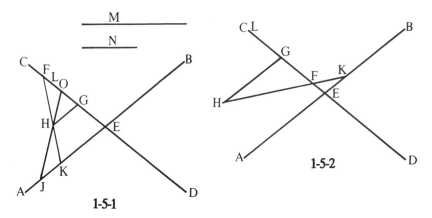

1-5-1

1-5-2

So point L also is a recipient. And line GL is a recipient in position. And since the ratio of KE to GF is like the ratio of HG to GL, then, by alternation, the ratio of KE to GH, that is, the ratio of EF to GF, is like the ratio of FG to GL. And by composition, the ratio of EG to GF is like the ratio of FL to LG. So rectangle EG by GL, which is equal to rectangle LF by FG, is a recipient. And it has been applied to a recipient line, namely, line LG, with the excess of a square. So point F is recipient in position. And it is known that it falls between points E & G. And point H is a recipient. So line HK is positioned.

And this problem will be synthesized thus: Let the parallel line be in the same state of parallelism. And let the given ratio be the ratio of M to N, which is the same as the ratio of HG to GL. And let there be applied to line LG a rectangle equal to rectangle EG by GL and exceeding by a square, namely, the rectangle LF by FG. And let line FH be joined and produced in a straight line. Then I say that line HK does what the problem requires. That is, that the ratio of KE to GF is like the ratio of M to N.

For, since the rectangle EG by GL is equal to the rectangle LF by FG, proportionally, the ratio of EG to GF is like the ratio FL to GL. And by separation, the ratio of EF to FG is like the ratio of FG to GL. That is, the ratio of KE to GH is like the ratio of FG to GL. And by alternation, the ratio of KE to FG is like the ratio of GH to GL. However the ratio of HG to GL is like the ratio of M to N. So the ratio of KE to GF is like the ratio of M to N. So line HK does what the problem requires.

Then I say that it alone does so. For if it were possible for another to do so, then let some other line, such as HO, be drawn. Then if the line HO cuts off a ratio the same as the ratio of M to N, the ratio of KE to FG is like the ratio of EO to GJ. However the ratio of EK to FG is like the ratio of HG to GL. So the ratio of EO to GJ is like the ratio of HG to GL. So by alternation, the ratio of EO to GH is like the ratio of JG to GL. However the ratio of EO to GH is like the ratio of EJ to JG. So the ratio of EJ to JG is like the ratio of JG to GL. And by composition, the ratio of EG to JG is like the ratio of JL to GL. So rectangle EG by GL is equal to rectangle LF by FG. But that is something impossible, because the rectangle LF by FG was made equal to rectangle EG by GL. So line HK alone does what the problem requires. And this line is a line cutting off increasing ratios as it is more distant from point E. This is clear from the fact that KH cuts off a ratio less than the ratios that line HO cuts off.

Case 3. Let a line HF be drawn in the third way, cutting off from lines EA, GD a ratio of EF to GK the same as a given ratio. So the ratio of EF to GK is a recipient. And let the ratio of HG to GL be the same as it. However the line GH is a recipient. So line GL is a recipient. But point G is a recipient. So point L also is a recipient. And line GL is in magnitude & position. And since the ratio of EF to GK is like the ratio of HG to GL, then, by alternation, the ratio of EF to GH is like the ratio of KG to GL. But the ratio of EF to GH is like the ratio of EK to KG. So the ratio of EK to KG is like the ratio of KG to GL. And by inversion and conversion, the ratio of KG to GE is like the ratio of GL to LK. So the rectangle LG by GE is equal to the rectangle GK by KL. However rectangle LG by GE is a recipient, because each of its lines is recipient. So rectangle GK by KL is a recipient. And it has been applied to the recipient line GL with the deficiency of a square. So point K is

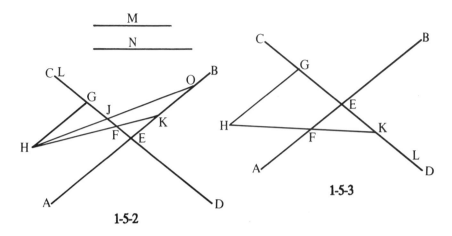

1-5-2

1-5-3

recipient. But point H is recipient. So line HK is recipient in position.

And since in the synthesis it is necessary to make the ratio of HG to GL the same as a given ratio, and to apply to line GL a rectangle equal to rectangle LG by GE and deficient by a square, namely, rectangle GK by KL, then, whenever the requisite line is drawn in the place of HK, we will say that HK is the line that does what the problem requires. But it is not always possible to draw the requisite line, because the rectangle LG by GE is sometimes greater than the square of half LG. Then the application is impossible. So for that reason the synthesis of the problem is not always possible for every domain <of the ratio>.

So, let it obtain in one way: by drawing the requisite line to the midpoint of line GL, at the place of point K. Then rectangle LG by GE is equal to rectangle GK by KL. So in this manner the problem can be done. But how will we find a ratio that <serves as> the ratio of HG to such a line as GL, while line GL is separated into halves at K, and while rectangle LG by GE is equal to rectangle rectangle GK by KL? This is like taking some point along line EG, such as point L, and separating GL into halves at point K, and requiring rectangle LG by GE to be equal to rectangle KG by KL.

Then let it be so. Now, since rectangle LG by EG is equal to rectangle GK by KL, the ratio of GL to LK is like the ratio of KG to GE. However line GL is double line LK. So line KG is double line GE. So line GE is equal to line EK. But line GE is recipient. So line EK is recipient in magnitude. But point E is a recipient. So point K is a recipient. But point G is a recipient. So line GK is a recipient, and it is equal to line KL. So line KL is recipient in magnitude. But point K is a recipient. So point L is a recipient. So the point sought for, which <determines> the ratio, is point L.

Then I say that, if HK be joined, the ratio of FE to GK is like the ratio of HG to GL. For, since line KE [is half of GK and GK is half of LG], the ratio of LG to GK is like the ratio of GK to KE, that is, the ratio of HG to EF. And by alternation, the ratio of HG to GL will be like the ratio of EF to GK. So, for the synthesis, it is required that line GE be set equal to line EK, and that HK be joined.

But let us first seek whether line HK cuts off a ratio less than, or greater than, all the lines which are drawn from point H cutting off lines from EA, GD. And that will be known in this way: The parallel line will be as before, and line EK will be made equal to line EG, and HK joined. It is necessary to find out out whether line HK cuts off a ratio of EF to GK greater than the ratio that the lines drawn from point H take off when they cut lines EA, GD. So line KL will be made equal to line GK. So rectangle LG by GE is equal to rectangle GK by KL. But the ratio of EF to GK is the same as the ratio of HG to GL, that is, the ratio of GH to four times GE. And let another line be drawn, such as HM. So it is necessary to compare the ratio of HG to GL with the ratio of NE to GM. And by alternation, it is necessary to compare the ratio of HG to EN with the ratio of LG to GM. But the ratio of GH to EN is like the ratio of GM to ME. So it is necessary to compare the ratio of GM to ME with the ratio of LG to MG. And by conversion, it will be necessary to compare the ratio of MG to GE with the ratio of GL to LM. And it will be necessary to compare the rectangle LG by GE with the rectangle GM by ML. However

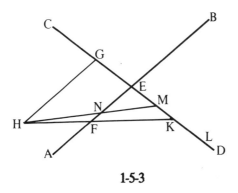

1-5-3

rectangle GK by KL is equal to rectangle LG by GE. So it is necessary to compare rectangle GK by KL with rectangle GM by ML. And the comparison is this: that rectangle GK by KL is greater than rectangle GM by ML, because the midpoint of line GL is point K.

Now, since rectangle GM by ML is less than rectangle GK by KL, and rectangle LG by GE is equal to rectangle GK by KL, rectangle GM by ML is less than rectangle LG by GE. So the ratio of MG to GE is less than the ratio of GL to LM. And by conversion, the ratio of GM to ME is greater than the ratio of LG to GM. However the ratio of GM to [ME] is like the ratio of GH to EN. So the ratio of GH to EN is greater than the ratio of LG to GM. And by alternation, the ratio of GH to GL is greater than the ratio of EN to GM. However the ratio of HG to GL is like the ratio the ratio of EF to GK. So the ratio of EF to GK is greater than the ratio of EN to GM. So line HK cuts off a ratio greater than the ratio that line HM cuts off. And likewise it is clear that it cuts off a ratio of EF to GK which is greater than all the ratios that are cut off by lines drawn from point H cutting off from the two lines EA, GD.

And I say that the lines nearer line HK always cut off ratios greater than the ratios that the lines more distant from it cut off. For, since the ratio of NE to GM is less than the ratio of EF to GK, while the ratio of EF to GK is like the ratio of

GH to GL, the ratio of NE to GM is less than the ratio of GH to GL. Then, if it be made that, as the ratio of NE to GM, so the ratio of HG to another line, that line will be greater than GL. Let it be GJ. So, just as in the analysis, it is clear that, with the ratio of NE to GM like the ratio of HG to GJ, the rectangle GJ by GE is equal to the rectangle GM by MJ. And so the ratio of NE to GM is the same as the ratio of HG to GJ.

So let another line such as HO be drawn. Then it will be necessary to compare the ratio of NE to MG with the ratio of PE to GO. However the ratio of NE to GM is like the ratio of HG to GJ. So by alternation, it will be necessary to compare the ratio of GH to EP with the ratio of JG to GO. However the ratio of GH to EP is like the ratio of GO to OE, [so it is necessary to compare GO to OE] with the ratio of JG to GO. And by conversion, it will be necessary to compare the ratio of OG to GE with the ratio of GJ to JO, and to compare the rectangle JG by GE with rectangle GO by OJ. However rectangle JG by GE is equal to rectangle GM by MJ. So it will be necessary to compare rectangle GM by MJ with rectangle GO by OJ, and also to compare rectangle GK by KJ with rectangle GM by MJ. However rectangle JG by GE is equal to rectangle GM by MJ. So it will be necessary to compare rectangle GK by KJ with rectangle JG by GE.

And the comparison is made like this: Since the rectangle GK by KL is equal to rectangle LG by EG, it is necessary to compare rectangle GK by JL with rectangle JL by GE, for they are the excesses over the two rectangles GK by [KL] & LG by GE. And the comparison is this: that rectangle JL by GK is greater than rectangle JL by GE. So for that reason rectangle GK by KJ is greater than rectangle JG by GE, which is equal to rectangle GM by MJ.

So, since rectangle GK by JK is greater than rectangle GM by MJ, and point O is known [to be more distant from K than M is], rectangle GM by MJ is greater than rectangle GO by OJ. But rectangle JG by GE is equal to rectangle GM by MJ. So the ratio of OG to GE is less than the ratio of GJ to JO. And by conversion, the ratio of GO [to OE] is greater than the ratio of JG to GO. And the ratio of OG to EO is like the ratio of GH to EP. So the ratio of HG to PE is greater than the ratio of JG to

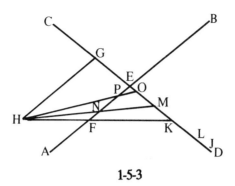

1-5-3

GO. And by alternation, the ratio of HG to GJ is [greater than the ratio of PE to GO. But the ratio of HG to GJ is] like the ratio of NE to GM. So the ratio of NE to GM is greater than the ratio of PE to GO. So for that reason line HM cuts off a ratio greater than the ratio that line HO cuts off. So it is clear that lines nearer to line HK cut off ratios greater than the ratios that lines farther from it cut off. And it is clear that the ratio of EF to GK was derived from <the particular instance that was supposed>, and it is the same as the ratio of GH to four times GE, because GL is four times GE.

And this problem will be synthesized thus: The parallel line will be as before. And let line EK be made equal to line GE. And let line HK be joined. So line HK will cut off a ratio greater than the ratios that the lines that are drawn from point H cut off when cutting off from lines EA, GD. So the given ratio is either the same as the ratio of EF to GK, which is the one the same as the ratio of HG to four times GE, for it is a special instance, as we have explained in the analysis; or it is greater than the ratio of EF to GK; or less than it.

Now when the given ratio is the same as the ratio of EF to GK, then the line HK does what the problem requires. And it will be clear that it alone does so, because lines nearer to it always cut off ratios greater than the ratios that lines farther away from it cut off.

And if it is greater than the ratio of EF to GK, the problem will be impossible, because the given ratio is greater than <the one exceeding the ratio of all the lines cut off.>

So let the ratio of M to N be less than the ratio of EF to GK, and let line GK be made equal to line KL. So rectangle LG by GE is equal to rectangle GK by KL. However the ratio of EF to GK is the same as the ratio of HG to GL. And since the ratio of M to N is less than the ratio of EF to GK, that is, less than the ratio of HG to GL, let the ratio of HG to some other line be made the same as the ratio of M to N. So that line is longer than line GL. Let it be line GJ. And since line KG is greater than line GE, rectangle JL by KG is greater than rectangle JL by EG. And since rectangle LG by GE is equal to rectangle GK by KL, the rectangle GK by KL will be annexed to rectangle GK by JL, and rectangle LG by GE will be annexed to rectangle JL by EG. So the whole rectangle KJ by GK is greater than the whole rectangle JG by GE. So it is possible to apply to line GJ a rectangle equal to rectangle JG by GE [and deficient by a square] in two ways, namely on both sides of point K. So let the two points of the required lines be points O & R, and let HR & HO be joined. Then I say that each of the two lines HO, HR does what the problem requires.

For, since rectangle JG by GE is equal to rectangle GO by OJ, the ratio of OG to GE is like the ratio of GJ to JO. And by conversion, the ratio of OG to OE is like the ratio of JG to OG. However the ratio of OG to OE is like ratio of HG to EP. So the ratio of HG to EP is like the ratio of GJ to GO. And by alternation, the ratio of HG to GJ is like the ratio of EP to GO. But the ratio of HG to GJ is like the ratio of M to N. So the ratio of M to N is like the ratio of EP to GO. So line HO will do what the problem requires. And similarly, it is clear that line HR does. So each of the two lines HO, HR will do what the problem requires.

And since we have made it clear <under what circumstances> the problem will be synthesized according to all the cases, let us clarify in how many ways the synthesis comes about for each of its domains. Let the parallel line be as before.

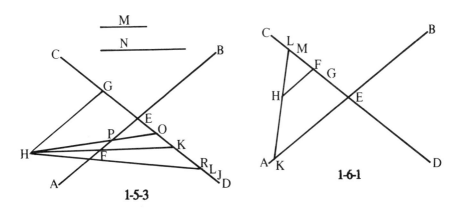

1-5-3   1-6-1

So the given ratio is either the same as the ratio of HG to four
times GE, or it is greater, or it is less. Now if it is the same,
then the problem will be synthesized according to three of the
cases: according to the first and second cases, and one time in
the third case. If the ratio is greater, then the problem will be
synthesized according to two cases: according to the first case
and the second. And if it is less, then it will be synthesized
according to four cases: according to the first case and the
second, and in two ways in the third.

Locus Six. And let the line parallel to line AB passing through
point H fall above point G, namely, line HF. That is, point G is
between F and E. The line drawn from point H is in one of
four cases. Either it cuts off a ratio from the two lines CG, EA;
or from CG, EB; or from GE, EB; or from GD, EA.

Case 1. So let a line KL be drawn in the first way, cutting off
from the two lines CG, EA a ratio of EK to GL the same as a
given ratio. And let the ratio of line HF to line GM be the
same as the ratio of EK to LG. But line HF is recipient. So line
GM is a recipient in position and in magnitude. But point G is
a recipient. So point M is a recipient. But the ratio of EK to GL
is like the ratio of HF to GM. And by alternation, the ratio of
EK to FH is like the ratio of LG to GM. However the ratio of
EK to FH is like the ratio of LE to LF. So the ratio of LE to
LF is like the ratio of LG to GM. And by separation, the ratio

of EF to FL is [like the ratio of LM to GM. So the rectangle EF by GM] is equal to the rectangle FL by LM. And the rectangle MG by FE is a recipient, because each of the two is recipient. So rectangle FL by LM is a recipient. And it has been applied to the recipient line FM with the excess of a square. So point L is a recipient. So line LK is positioned.

And this problem will be synthesized thus: With the parallel line remaining the same, let the given ratio be the ratio of N to J. And let the ratio of GM to HF be made the same as the ratio of J to N. And let there be applied to line FM a rectangle equal to rectangle GM by FE and exceeding by a square, namely, the rectangle FL by LM. And let LH be joined and produced in a straight line. Then I say that the line LK will do what the problem requires. That is, that the ratio of N to J is like the ratio of EK to GL.

For, since rectangle EF by GM is equal to rectangle FL by LM, the ratio of EF to FL is like the ratio of LM to GM. And by composition, the ratio of EL to LF, that is, the ratio of EK to FH, is like the ratio of LG to GM. And by alternation, the ratio of EK to GL is like the ratio of HF to GM. But the ratio of HF to GM is like the ratio of N to J. So the ratio of EK to GL is like the ratio of N to J. So line LK does what the problem requires.

And I say that it alone does so. For if it were possible, let there be another, such as OP. Then, if line OP cuts off a ratio the same as the ratio of N to J, the ratio of KE to GL is like the ratio of EP to GO. But that is impossible, because the latter is less than the former. So it is clear that the lines near point E, such as line OP, cut off a ratio less than the ratio that the lines distant from it cut off.

Case 2. With the parallel line being the same as before, let a line HL be drawn in the second way, cutting off from the lines CG, EB a ratio of LE to GK the same as a given ratio. And let the ratio of FH to GM be made the same as the ratio of LE to GK. But line FH is a recipient. So line GM also is recipient in position and in magnitude. So point M is recipient. But point F

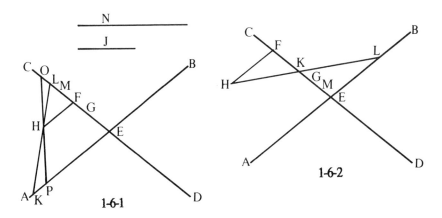

is recipient. And so line FM is recipient in position and in magnitude. And since the ratio of LE to GK is like the ratio of HF to GM, then, by alternation, the ratio of LE to FH is like the ratio of KG to GM. And the ratio of LE to FH is like the ratio of EK to KF. And by composition, the ratio of EF to FK is like the ratio of KM to MG. And rectangle FE by GM is a recipient, because each of them is a recipient. So rectangle MK by KF is a recipient. And it has been applied to line FM with the deficiency of a square. So point K is a recipient. But point H is a recipient. So line HL is positioned.

And since it is necessary for the synthesis to make the ratio of FH to GM the same as the given ratio, and to apply to line FM a rectangle equal to rectangle FE by GM and deficient by a square, and since point K is on line FM, the requisite line is not always possible, because sometimes rectangle FE by GM is greater than the square on half of line FM. Then the application is impossible. So for that reason it is impossible for the problem always to be synthesized in every way.

So first let one case obtain: namely, if the requisite line is drawn to point K, which is at the middle of FM. But rectangle FK by KM is equal to rectangle FE by GM. So the problem can be done in this way. So we want to find a ratio such that, after we have made the ratio of FH to GM the same as it, and have cut line FM in half at K, the rectangle FK by KM will then be equal to rectangle EF by GM. This, then, amounts to knowing a point on GD like point M, and cutting line FM in half as though at point K, and setting rectangle FE by GM equal to rectangle FK by KM. So the ratio of EF to FK is like the ratio of KM to MG. And by conversion, the ratio of FE to EK is like the ratio of MK to KG. But line MK is equal to line KF. So the ratio of FE to EK is like the ratio of FK to KG. And the ratio of KE to EG, <will remain in place of one of the ratios> So the ratio of FE to EK is like the ratio of KE to EG. So line EK is a mean proportional of the two lines FE, EG. But each of FE, EG is a recipient. So line EK is a recipient in position and in magnitude. But point E is a recipient. So point K is a recipient. But point F also is a recipient. So line FK is recipient. And it is equal to line KM. So line KM is a recipient. But point K is a recipient. So point M also is a recipient.

And this analysis will be synthesized thus: Let there be taken a mean proportional of the two lines FE, EG, namely, line EK. Now, it is clear that line FK is greater than line KG, because the ratio of FE to EK is like the ratio of KE to EG, and so the ratio of FK to KG <will remain in place of one of them as remainder> So line FK is greater than line KG <by a remainder similar to that in one of the ratios> So let it be equal to line KM. So point M is the one sought for. That is, the rectangle FE by GM is equal to rectangle MK by KF.

For, since the ratio of FE to EK is like the ratio of KE to GE, but the ratio of the remainder to the remainder is like one of the ratios, then the ratio of FE to EK is like the ratio of FK to KG. But line FK is equal to KM. So the ratio of EF to EK is like the ratio of MK to KG. And by conversion, the ratio of EF to KF is like the ratio of KM to MG. So rectangle FE by MG is equal to rectangle MK by KF.

And let HK be joined and produced in a straight line. Then I say that line HL will do what the problem requires. That is, that the ratio of LE to GK is like the ratio of FH to GM. For, since rectangle FE by GM is equal to rectangle MK by KF, the ratio of FE to KF is like the ratio of KM to MG. And by separation, the ratio of EK to KF is like the ratio of KG to GM. However the ratio of EK to KF is like the ratio of LE to FH.

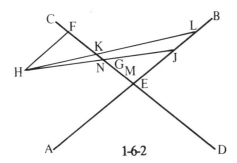

1-6-2

So the ratio of LE to FH is like the ratio of KG to GM. And by alternation, the ratio of LE to KG is like the ratio of FH to GM. So before the synthesis, it is necessary to take a mean proportional line between the two lines FE, EG, namely, line EK. Then HK will be joined and produced.

But we first have to find out whether line HL cuts off a ratio, namely, the ratio of LE to GK, greater or less than all the lines that are drawn from point H and cut off from the two lines EB, GC. And we find this problem thus: With the parallel line remaining the same, let there be taken a mean proportional line between the lines FE, EG, namely, line EK. And let HK be joined and produced in a straight line. It is necessary for us to seek out if line HL cuts off a ratio of LE to GK less than, or greater than, that which the lines drawn from point H cut off when they cut lines EB, GC. Then a line equal to line FK be marked off, namely, KM. So rectangle FE by MG is equal to rectangle MK by KF.

And let a line such as HN be drawn. So it is necessary to compare the ratio of JE to GN with the ratio of LE to GK. However the ratio of LE to GK is like the ratio of FH to GM. So it will be necessary to compare the ratio of JE to GN with the ratio of FH to MG. And by alternation, it will be necessary to compare the ratio of JE to FH with the ratio of GN to MG. However JE to FH is like the ratio of EN to NF. So it will be necessary to compare the ratio of EN to NF with the ratio of NG to GM. And by composition, it will be necessary to compare the ratio of EF to FN with the ratio of NM to MG. So it will be necessary to compare rectangle FE by MG with rectangle MN by NF. Now we find that its comparison is this: Rectangle MK by KF is greater than the rectangle MN by NF, since point K cuts line FM in half. So, since rectangle MK by KF is greater than rectangle MN by NF, while rectangle EF by MG is equal to rectangle MK by KF, so rectangle EF by MG is greater than rectangle MN by NF.

So the ratio of EF to NF is greater than the ratio of NM to MG. So by separation, the ratio of EN to NF is greater than the ratio of NG to GM. However the ratio of EN to NF is like the ratio of JE to FH. So the ratio of EJ to FH is greater than the ratio of NG to GM. And by alternation, the ratio of JE to NG is greater than the ratio of FH to MG. However the ratio of FH to GM is like the ratio of LE to GK. So the ratio of JE to NG is greater than the ratio of LE to GK. So for that reason line HL cuts off a ratio less than the ratio that HJ cuts off. And likewise it is clear that it cuts off a ratio less than the ratios that all the lines drawn from point H cut off, while cutting the two lines GC, EB.

And I say that the lines near line HL cut off ratios less than the ratios that the lines distant from it cut off. For, since the ratio of JE to GN is greater than the ratio of LE to GK, that is, the ratio of FH to GM, if the ratio of FH to some line is made the same as the ratio of JE to GN, then that line will be less than line GM. So let it be line GO. Likewise it is clear that rectangle EF by GO is equal to rectangle ON by NF. However the ratio of JE to GN is the same as the ratio of FH to GO.

Let another line, such as HX be drawn. So it is necessary to compare the ratio of XE to GP with the ratio of JE to GN. However the ratio of JE to NG is like the ratio of FH to GO. So it will be necessary to compare the ratio of XE to GP with the ratio of FH to GO. And by alternation, it will be necessary to compare the ratio of XE to FH with the ratio of PG to GO. However the ratio of XE to FH is like the ratio of EP to PF. And by composition, it will be necessary to compare the ratio

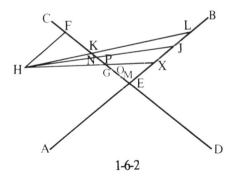

1-6-2

of FE to FP with the ratio of PO to GO. And it will be necessary to compare rectangle FE by GO with rectangle OP by PF. However rectangle ON by NF is equal to rectangle FE by OG. So it will be necessary to compare rectangle ON by NF with rectangle OP by PF. For that purpose it is necessary to compare rectangle OK by KF with rectangle ON by NF.

However rectangle FE by OG is equal to rectangle ON by NF. So when the rectangle OK by KF is compared with the rectangle EF by OG, the rectangle MK by KF is equal to rectangle FE by MG. Then, by subtracting from rectangle EF by MG the rectangle EF by OG, and from rectangle MK by KF the rectangle OK by KF, it will be necessary to compare rectangle MO by EF with rectangle MO by KF. And it is found that rectangle MO by EF is greater than MO by KF. So, since rectangle FE by MG is equal to rectangle MK by KF, and rectangle MO by EF is greater than rectangle MO by KF, rectangle OG by EF will remain less than rectangle OK by KF. However rectangle ON by NF is equal to rectangle OG by FE. So rectangle ON by NF is less than rectangle OK by KF.

And for that reason rectangle ON by NF is greater than rectangle OP by PF. However rectangle OG by FE is equal to rectangle ON by NF. So rectangle OG by FE is greater than rectangle OP by PF. So the ratio of FE to FP is greater than the ratio of PO to GO. And by separation, the ratio of EP to PF, that is, the ratio of XE to FH, will be greater than the ratio of PG to GO. And by alternation, the ratio of XE to PG will be greater than the ratio of FH to GO. However the ratio of FH to GO is like the ratio of JE to GN, and the ratio of XE to GP is greater than the ratio of JE to GN. So the line HJ cuts off a ratio less than the ratio that the line XH cuts off, and line HJ is nearer to line HL than line HX. So it is clear that the lines nearer line HL always cut off ratios less than the ratios that the lines farther from it cut off.

And this problem will be synthesized thus: With the parallel line remaining the same, let there be taken a mean proportional line between the two lines FE, EG, namely, line EK. And let line HK be joined and produced in a straight line. So line HL cuts off a ratio of LE to GK. So the given ratio is either the same as the ratio of LE to GK, or it is less, or greater.

Now, if it is the same as the ratio of LE to GK, LH makes the problem work. And it is clear that it alone does so.

And if it is less than LE to GK, then the problem will be impossible.

So let the given ratio be the ratio of N to J, which is greater than LE to GK. And FK will be equal to KM. So EF by GM is equal to MK by KF. And the ratio of LE to GK is the same as the ratio of FH to GM. And N to J is greater than LE to GK, and also greater than FH to MG. And it is known that, if N to J is the same as FH to another line, that line will be less than GM. Let it be GO. And since EF by MG is equal to MK by KF, and MO by EF is greater than MO by KF, OG by EF will remain less than OK by KF. And for that reason, it will be possible to apply to FO a rectangle equal to rectangle EF by OG and deficient from the completion of the line by a square, in two ways, namely, on both sides of K. [Let the two points of this application be points P and S.] Let HP, HS both be joined and produced. Then I say that each one of HX, HR will make the problem work, and the ratio of N to J is the same as XE to GP and the same as RE to GS.

For, FE by GO is equal to OS by SF and equal to OP by PF. So the ratio of one of them, say, the ratio of EF to FS, is like the ratio of SO to OG. And by separation, the ratio of ES to SF is the same as SG to GO. And the ratio of ES to SF is like the ratio of RE to FH. So the ratio of RE to FH is the same as GS

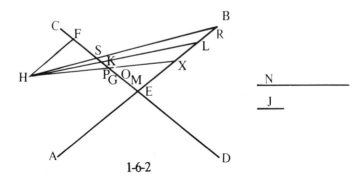

1-6-2

to GO. And by alternation, the ratio of RE to GS is the same as FH to GO. But FH to GO is the same as N to J. So the ratio of RE to GS is the same as N to J, and each one of HX, HR will make the problem work. And it is clear that only they do so, and that the lines nearer HL cut off ratios less than the ratios that the lines farther away cut off.

The restriction of the ratio will be recognized thus: The ratio of LE to GK is same as FH to GM. But GM is the excess of FE, EG over FE, EM. And FE, EM is equal to twice KE, because FK is equal to KM. And double KE equals in square FE by four times EG, because EK is a mean proportional between the two lines FE, EG. So GM is the excess of FE, EG over the line that equals in square FE by four times EG. So the ratio of LE to KG, [which was the least of all the ratios] that lines drawn from H cut off when they cut EB, GC, is the same as the ratio of FH to the excess of FE, EG over the line that equals in square four times FE by EG. And that is what we wanted to show.

Case 3. With the parallel line remaining the same as before, let a line HL be drawn according to the third case, cutting off from the two lines GE, EB a ratio of LE to KG the same as a given ratio. And let it be made that, as the ratio of LE to KG, so the ratio of same as FH to GM. But FH is a recipient. So GM is a recipient in position and in magnitude. But point G is positioned. So M is positioned. But F is positioned. So line FM is positioned. And since LE to KG is the same as FH to GM, then, by alternation, LE to FH, that is, EK to KF, is the same as KG to GM. And by composition, FE to FK is the same as KM to MG. So EF by GM is equal to FK by KM. And EF by GM is recipient, because each one of EF, GM is recipient. So FK by KM is recipient. And it has been applied to a recipient line FM with the excess of a square. So point K is recipient. But point H is recipient. So line KL is positioned.

And this problem will be synthesized thus: The parallel line will remain the same. And let the given ratio be the ratio of N to J. Let the ratio of FH to GM be made the same as it. And let there be applied to line FM a rectangle equal to rectangle FE by GM exceeding the line by a square. Now it cannot be FG by GM, because line EF is greater than FG. Nor can it be FE by EM, because line EM is greater than line MG. So it is clear that point K will cut off between G and E. So let it be FK by KM. And let HK be joined and produced in a straight line. Then I say that HL will make the problem work. That is, that LE to KG is the same as N to J.

For, since EF by GM is the same as FK by KM, the ratio of FE to KF is the same as KM to MG. And by separation, the ratio of EK to KF, that is, LE to FH, is the same as KG to GM. And by alternation, the ratio of LE to KG is the same as FH to GM, which is the same as N to J. So the ratio of LE to KG is the same as N to J. So line HL makes the problem work.

And I say that it alone does so. For, if it were possible for some other to do so, then let another such as HP be drawn, cutting off a ratio the same as N to J. So PE to OG is the same as LE to KG. But this is something impossible, because the antecedent is less than the antecedent, and the consequent is greater than the consequent. So it is clear that HP cuts off a ratio less than HL, because the latter is excessive, as we have made clear.

Case 4. With the parallel line the same as before, let a line HL be drawn according to the fourth case, cutting off from EA, GD a ratio of EK to LG the same as a given ratio, while the ratio

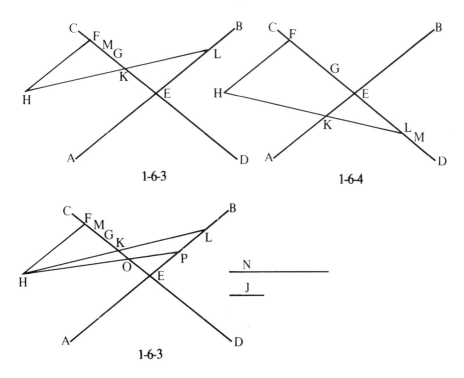

1-6-3

1-6-4

1-6-3

of FH to GM is made the same as it. But FH is a recipient. So MG is a recipient in magnitude and in position. But point G is a recipient. So point M is a recipient. And since the ratio of FH to GM is like the ratio of EK to GL, then, by alternation, the ratio of FH to EK, that is, the ratio of FL to LE, is like the ratio of MG to GL. And by conversion, the ratio of LF to FE is the same as MG to ML. So the rectangle MG by FE is equal to the rectangle FL by LM. But rectangle MG by FE is a recipient, because each one of FE, MG is a recipient. And it has been applied to a recipient line, namely, FM, deficient by a square. So point L is a recipient. But point H is a recipient. So line HL is a recipient in position.

And since in the synthesis it is necessary to make the ratio of FH to GM the same as the given ratio, and to apply to line FM a rectangle equal to rectangle FE by GM and deficient

from the completion of the line by a square, namely, rectangle
FL by LM, and to join line HL: so it will be necessary to take
the point L. But that is not always possible in every way. And
for that reason, the synthesis of the problem will not be
possible according to each of its domains.

However, there is one case that will make the proportion
work: cutting off half of FM at point L. And that results in a
problem like this one: How shall we find a ratio when it has
been made the same as the ratio of FH to GM, and when FM is
cut in half, with the rectangle FL by LM being set equal to the
rectangle GM by FE? This means taking something like GM on
<GO> and cutting FM in half at point L, while setting
rectangle FL by LM equal to GM by FE. Then, since rectangle
FL by LM is equal to GM by FE, the ratio of LF to FE is the
same as GM to ML. And by separation, the ratio of GL to LM
is the same as LE to EF. But line ML is equal to line LF. So the
ratio of GL to LF is the same as LE to FE, and the ratio of GE
to EL will remain like a ratio the same as LE to EF. So line EL
is a mean proportional between FE, EG. And each one of FE,
EG is a recipient. So EL is a recipient in position. But point E
is a recipient. So point L is a recipient. But point F also. So line
FL is a recipient. And it is equal to LM. So line LM is a
recipient in position. But point L is a recipient. So is the point
sought for, namely, point M.

And the analysis will be synthesized thus: With the parallel
line remaining the same, let there be taken a mean proportional
line between the two lines FE and EG, namely, EL, and let LM
be made equal to FL. Then I say that the point sought for is
point M, and that rectangle FL by LM is equal to rectangle GM
by FE.

For, since between FE, EG there is a mean proportional
line, namely, EL, the ratio of GE to EL is like the ratio of EL
to FE. And as a ratio like one of the ratios so a ratio like the
sums of the ratios. So the ratio of LG to LF is like the ratio of
LE to EF. But FL is equal to LM. So GL to LM is the same as
LE to EF. And by composition, the ratio of GM to ML is like
the ratio of LF to EF. So rectangle FL by LM is equal to
rectangle GM by FE. So the point sought for is point M.

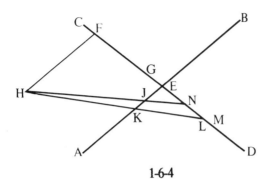

1-6-4

And let HL be joined. Then I say that EK to GL is the same as FH to GM, because GM to ML is the same as LF to FE. And by conversion, MG to GL is the same as FL to LE, that is, FH to EK. And by alternation, the ratio of FH to GM is the same as EK to GL. So prior to synthesizing, it is necessary to take a mean proportional line between the two lines FE, EG, and then to join HL.

But first, does line HL cut off a ratio of EK to GL less than, or greater than, all the lines that are drawn from point H and cut off from EA, GD? And we find this problem thus: With the parallel line remaining the same as before, let there be taken a mean proportional line between FE & EG, namely, EL. And let HL be joined. We will first inquire whether HL cuts off a ratio of EK to GL greater or less than all the lines that are drawn from point H and cut GD, EA. Let LM be equal to FL. So rectangle FL by LM is equal to GM by FE. And the ratio of EK to GL is the same as FH to GM.

Now let another line be drawn, such as HN. So it will be necessary to correlate EK to GL with EJ to GN. But the ratio of EK to GL is the same as FH to GM. So FH to GM is to be correlated with EJ to GN. And by alternation, FH to JE is to be correlated with GM to GN. And the ratio of FH to JE is the same as FN to EN. So FN to EN is to be correlated with GM to GN. Then, by conversion, FN to FE is to be correlated with GM to NM, and rectangle GM by FE with FN by NM. But rectangle GM by FE is equal to rectangle FL by LM. So it will be necessary to correlate rectangle FL by LM with rectangle FN by NM. Now its relation is this: that FL by LM is greater than FN by NM, because point L bisects FM.

And since FN by NM is less than FE by GM, the ratio of FN to FE is less than GM to MN. And by conversion, line FN to NE, that is, FH to EJ is greater than MG to GN. And by alternation, the ratio of FH to GM, that is, EK to GL, is greater than EJ to GN. So line HL cuts off a ratio greater than HN. And likewise, it is clear that it cuts off a ratio greater than the ratios that the lines departing from H cut off.

So line HL cuts off a ratio greater than the ratio of EJ to GN. And among all the lines that are drawn from point H and cut GD, EA, I say that the line more proximate to HL cuts off a ratio greater than that which the one more remote from it cuts off. For, since the ratio of JE to GN is less than the ratio of EK to GL, that is, FH to GM, when it is made that, as the ratio of EJ to GN, so the ratio of FH to another line greater than GM, then let it be GO. And likewise, it can be made clear, just as was done in the analysis, that rectangle GO by FE is equal to rectangle FN by NO.

Now let another line such as HP be drawn. Then it will be necessary to correlate the ratio of EJ to GN with the ratio of EX to GP. However the ratio of EJ to GN is like the ratio of FH to GO. So it will be necessary to correlate the ratio of FH to GO with the ratio of EX to GP. And by alternation, it will be necessary to correlate FH to EX, that is, FP to PE, with OG to GP. And by conversion, it will be necessary to correlate the ratio of FP to FE with the ratio of GO to OP. And it will be necessary to correlate rectangle GO by FE with rectangle FP by PO. But rectangle GO by FE is equal to rectangle FN by NO. So it will be necessary to correlate rectangle FN by NO with rectangle FP by PO. And for that purpose, it will be necessary to correlate rectangle FL by LO with rectangle FN by NO. But rectangle FN by NO is equal to GO by FE. So it will be necessary to correlate rectangle FL by LO with rectangle GO by FE.

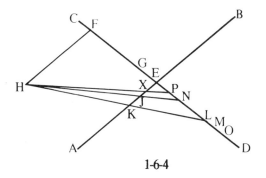

1-6-4

And we will find its relation to be such as we will now explain: Rectangle FL by LM is equal to rectangle FE by GM. But FL by LM is deficient from rectangle FL by LO, and rectangle FE by GM is deficient from rectangle FE by GO. And the remainder is to be correlated with the remainder. So it will be necessary to correlate rectangle FL by MO with rectangle FE by MO. And its relation is this: that it is greater. That is, rectangle FL by MO is greater than rectangle FE by MO, because LF is greater than FE.

And because rectangle FE by MO is less than rectangle FL by LM, the whole rectangle FE by GO is less than the whole rectangle FL by LO. But rectangle FE by GO is equal to rectangle FN by NO. So rectangle ON by NF is less than rectangle FL by LO. And for that reason, rectangle FP by PO will be less than rectangle FN by NO. However rectangle FN by NO is equal to rectangle FE by GO. So rectangle FP by PO is less than the rectangle FE by GO. So the ratio of PF to FE is less than the ratio of GO to OP. And by conversion, the ratio of FP to PE, that is, the ratio of FH to EX, is greater than the ratio of OG to GP. And by alternation, the ratio of FH to GO is greater than the ratio of EX to GP. And the ratio of FH to GO is like the ratio of EJ to GN. So the ratio of EJ to GN is greater than the ratio of EX to GP. So the more proximate line IIN cuts off a ratio greater than the ratio that line HP cuts off. And for that reason it is clear that the more proximate line always cuts off a ratio greater than the ratio that the farther one cuts off.

And the problem will be synthesized thus: With the parallel line remaining the same, let there be taken a mean proportional of the two lines FE, EG, namely, EL. And let HL be joined, cutting off the ratio of EK to LG. So the given ratio will either be the same as EK to GL, or less than it, or greater.

Then if it is the same as EK to GL, HL makes the problem work. And it is clear that it alone does so.

And if the given ratio is greater than EK to GL, then the problem cannot be found, because HL cuts off a ratio of EK to GL greater than each of the lines that are drawn from point H cutting off from GL, EA.

So let the ratio of N to J be less than the ratio of EK to GL. And let line LM be made equal to line LF. So rectangle GM by FE is equal to rectangle FL by LM. But the ratio of EK to GL is like the ratio of FH to GM. And since the ratio of N to J is less than the ratio of EK to GL, that is, FH to GM, when it is made that, as the ratio of N to J, so the ratio of FH to another line greater than GM, then let that line be GO. Then, since rectangle OM by LF is greater than OM by EF, and rectangle ML by LF is equal to rectangle FE by MG, the whole rectangle LF by LO is greater than the whole rectangle FE by GO. So rectangle FE by GO will be able to be applied to line OF deficient by a square, in two ways, namely, on each side of point L. So let the two points be points S, P be cited. And let HS, HP be joined. Then I say that each one of HS, HP will make the problem work. That is, that the ratio of ER to GS is the same as N to J, and the ratio of EX to GP is like the ratio of N to J.

For, since rectangle FS by SO is equal to FE by GO, the ratio of SF to FE is the same as the ratio of GO to OS. And by conversion, the ratio of FS to SE, that is, the ratio of FH to ER, is the same as GO to GS. And so by alternation, the ratio of FH to GO is the same as the ratio of ER to GS. But the ratio of FH to GO is like the ratio of N to J. So the ratio of N to J is the same as ER to GS. And likewise it is clear that the ratio of N to J is the same as EX to GP. And each one of the two lines HS, HP will make the problem work.

And it is clear that they alone do so, and that the line more

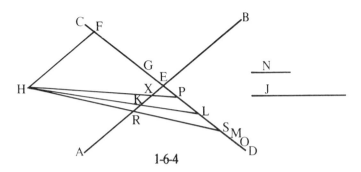

1-6-4

proximate to HL will cut off a ratio greater than the one farther from it.

And the limit of the ratio will be made known thus: Since the ratio of EK to GL is the same as the ratio of FH to MG; and line GM is GL, LM together, that is, FL, LG together, because line ML is equal to LF; and since FL, LG together are equal to FE, EG and the double of EL, which is itself equal in square to four times the rectangle FE by EG: so the ratio of EK to LG is the same as the ratio of FH to the line composed from FE, EG and the line equaling in square four times the rectangle FE by EG.

We have made clear how the problem will be synthesized according to each of its cases. It remains for us to explain in how many ways the problem will be synthesized for all its domains. With the parallel line remaining the same as before, let there be taken a mean proportional line between FE, EG and let this mean be either EK or EN. And let HN & HL be joined. And mark off KM equal to FK, and NJ equal to FN. Now, the least of the ratios according to the second case is the ratio of LE to GK, that is, the ratio of HF to GM. And the greatest of the ratios according to the fourth case is the ratio of ER to GN, that is, HF to GJ. And since the smallest ratio according to the second case is the ratio of HF to GM, and the largest according to the fourth case is the ratio of FH to GJ, and it is clear that FH to GM is larger than FH to GJ: so the given ratio is either the same as the ratio of FH to GM, or less than the ratio of FH to GM but greater than the ratio of FH to GJ, or greater than the ratio of FH to GM, or the same as the ratio of FH to GJ, or less than the ratio of FH to GJ.

Now, if it is the same as the ratio of FH to GM, it obtains according to three cases: according to the first and third case, because they each cut off any ratio; and once according the second case. It does not obtain according to the fourth case, because the ratio of FH to GM is larger than FH to GJ.

If it be less than FH to GM but larger than FH to GJ, the problem will obtain according to two ways: the first and the third. It will not obtain according to the second nor the fourth, because it was given less than the least and larger than the largest.

If it be given greater than FH to GM, then the problem will obtain according to four cases: according to the first and third, and in both ways of the second case. It will not obtain according to the fourth case, because it was given larger than FH to GJ.

If it be given the same as FH to GJ, it will obtain according to three cases: according to the first and the third and once in the fourth. And it will not obtain according to the second, because the ratio of FH to GJ is less than FH to GM.

And if it be less than FH to GJ, it will obtain according to four cases: the first, the third, and twice in the fourth. But it will not obtain according to the second, because it was given less.

And we have been able to explain it according to the variety of ratios.

Locus Seven. Let the recipient point be as before, but let the parallel line cut below point G, that is, between it and point E, as line HF. Then the lines drawn from point H obtain in four ways. Either they are drawn cutting off a ratio from GC, EA; or from GE, EA; or from GE, EB; or from GD, EA.

Case 1. So let KHL be drawn in the first way, cutting off from the two lines GC, EA a ratio of EL to GK the same as a given ratio. And as the ratio of EL to GK, so the ratio of FH to GM.

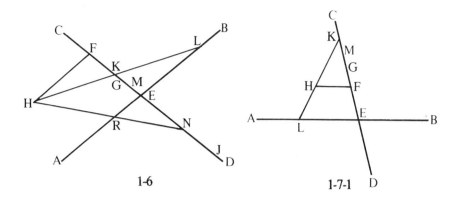

1-6   1-7-1 D

But FH is recipient. So GM is recipient in position and in magnitude. But point G is positioned. So M is recipient. And since the ratio of EL to GK is the same as FH to GM, then, by alternation, EL to FH, that is, EK to KF, will be the same as KG to GM. And by separation, EF to FK will be the same as KM to MG. So that which is contained by EF & GM is equal to FK by KM. However EF by GM is recipient, because each one of the two is recipient. So FK by KM is recipient. And it has been applied to the recipient line FM with the excess of a square. So point K is recipient. But point H is recipient. So line KL is positioned.

And this problem will be synthesized thus: Let the parallel line be as before. And let the given ratio be the ratio of N to J. And let the ratio of FH to GM be like it. And let there be applied to line FM a rectangle equal to rectangle EF by GM with an excess of a square over the completion of the line. So let it be rectangle FK by KM. And let KH be joined and produced. Then I say that KL will make the problem work. That is, that the ratio of N to J is the same as EL to GK.

For, the rectangle EF by GM is equal to the rectangle FK by KM. So the ratio of EF to FK is the same as KM to MG. And by composition, the ratio of EK to KF, that is, EL to FH, is the same as KG to GM. And by alternation, the ratio of EL to KG is the same as FH to GM, that is, N to J. So line KL will make the problem work.

And I say that it alone does so. For, if it were possible for one other than it to make the problem work, then let another line be drawn, such as OHP. So the ratio of OE to GP is like the ratio of EL to GK. But that is impossible, because the antecedent is greater than the antecedent while the consequent is less than the consequent. And from that it is clear that OP is an excessive line, which cuts off a ratio greater than what KL cuts off.

Case 2. Let the parallel line be as before. And let KL be drawn in the second way, cutting off from the two lines GE, EA a ratio of EL to KG the same as a given ratio. And let the ratio of HF to GM be the same as the ratio of EL to KG. And since the ratio of EL to KG is the same as FH to GM, and by alternation, the ratio of EL to FH, that is, EK to KF, is the same as KG to GM, while line EK is longer than line KF, so line KG is longer than GM. But line GM is a recipient in position and in magnitude. And point G is a recipient. So point M is a recipient. And since the ratio of EK to KF is like the ratio of KG to GM, then, by separation, the ratio of EF to FK is like the ratio of KM to MG. So rectangle EF by GM is equal to rectangle KF by KM. And rectangle EF by GM is a recipient, because each of the two is a recipient. So rectangle FK by KM is a recipient. And it has been applied to the recipient line FM, deficient by a square. So point K is recipient. But point H is recipient. So line KL is a recipient.

And since in the synthesis the ratio of FH to GM has to be made the same as a certain ratio, and there must be applied to line FM a rectangle like rectangle FK by KM equal to rectangle FE by GM and deficient by a square; and since point K is on FM, and the relation of the rectangle <to the square on half of FM> is not known: so for these reasons we cannot always synthesize the problem in every way.

So first let it obtain in one way, namely, when point K is at the midpoint of FM. Then there will be a problem like this one: How will we find a ratio the same as the ratio of FH to GM, when MF is cut in half [at K], and while rectangle EF by GM is

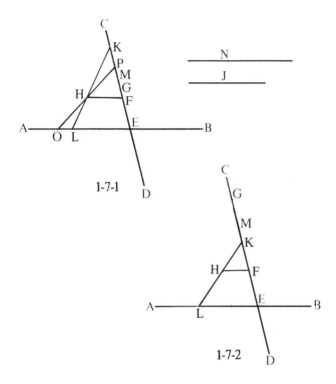

1-7-1

1-7-2

equal to FK by KM? Finding it amounts to taking a point on FG like point M, and cutting MF in half at point K, while rectangle EF by MG is set equal to rectangle KF by KM. So let it be found.   Then, since rectangle EF by MG is equal to rectangle FK by KM, the ratio of GM to MK is the same as the ratio of KF to FE. And by composition, the ratio of GK to KM, that is, GK to KF, is like the ratio of KE to FE. And the whole to the whole <is in the place of one> of the ratios. So the ratio of EG to EK is like the ratio of KE to EF. So line EK is a mean proportional between the two lines GE, EF. And each one of GE, EF is a recipient. So line EK is a recipient in position. But point E is positioned. So point K is a recipient. But point F is a recipient in position. So line KF is a recipient. And it is equal to line KM. So line KM is a recipient in position and in magnitude. But point K is a recipient. So point M is a recipient. And the point sought for is point M.

And this analysis will be synthesized thus: With the parallel line being as before, let there be taken a mean proportional line between lines GE & EF, namely, line EK. And let line KM be made equal to line FK. Now, M falls within points E & G, because line GK is greater than line KF. And that is so because the ratio of GE to EK is like the ratio of EK to EF, and <either of the ratios> is the same as the remainder to the remainder; that is, the ratio of GE to EK is the same as GK to KF. However the antecedent is greater than the consequent. So line GK is greater than KF. And let KH be joined and produced in a straight line. Then I say that rectangle EF by MG is equal to rectangle FK by KM, and that the ratio of EL to KG is the same as the ratio of FH to GM.

For, since line KE is a mean proportional between the two lines GE & EF, the ratio of GE to EK is like the ratio of EK to EF. And the remainder GK is to the remainder KF, that is, to KM, like one of the ratios, that is, the ratio of KE to EF. So by separation, the ratio of GM to MK is like the ratio of KF to FE. So rectangle EF by GM is equal to rectangle MK by KF. And also, since the ratio of KE to EF is like the ratio of GK to KM, and by conversion, the ratio of EK to KF, that is, the ratio of EL to FH, is like the ratio of KG to GM, then, by alternation, the ratio of EL to KG is like the ratio of FH to GM. So prior to the synthesis it will be necessary to take a mean proportional line between the two lines GE & EF, namely, line EK, and to join line KHL.

Now we seek whether line KHL cuts off a ratio, that is, the ratio of EL to KG, greater or less than all the lines drawn from H cut off from EA, GE. And this problem will be found as follows: Let the parallel line remain same. And let there be taken a mean proportional line between the two lines GE & EF, namely, line EK. And let line KHL be joined. And it will be necessary to seek whether line KL cuts off a ratio of EL to KG larger or smaller than all the lines that are drawn from point H and cut off from the two lines GE, EA. Let the line MK be made equal to the line FK. So rectangle EF by GM is equal to rectangle FK by KM, and the ratio of LE to KG is the same as the ratio of FH to GM.

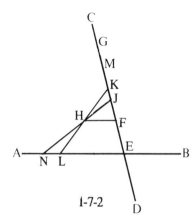

1-7-2

And let another line be drawn, such as NJ. So it will be necessary to compare the ratio of NE to GJ with the ratio of LE to KG, that is, FH to GM. And by alternation, it will be necessary to compare the ratio of EN to FH, that is, the ratio of EJ to JF, with the ratio of JG to GM. And by separation, to compare the ratio of EF to FJ with the ratio of JM to GM. And it will be necessary to compare rectangle FE by MG with rectangle FJ by JM. However rectangle FE by MG is equal to rectangle FK by KM. So it will be necessary to compare rectangle FK by KM with rectangle FJ by JM. But its relation is this: that rectangle FK by KM is greater than rectangle FJ by JM, because line FK is equal to line KM.

And because rectangle FK by KM is greater than rectangle FJ by JM, while rectangle FK by KM is equal to rectangle FE by GM, so rectangle EF by GM is greater than rectangle FJ by JM. So the ratio of EF to JF is greater than the ratio of JM to GM. And by composition, the ratio of EJ to JF, that is, the ratio of EN to FH, is greater than the ratio of GJ to GM. And by alternation, the ratio of EN to JG is greater than the ratio of FH to GM, that is, the ratio of EL to KG. And for that reason, line KL cuts off a ratio less than the ratio that line NJ cuts off. And likewise, it is clear that, among all the lines that are drawn from point H cutting off FG & EA, line KL cuts off the least of the ratios.

And since the ratio of EL to KG, that is, the ratio of FH to GM, is less than the ratio of EN to GJ, then, when it has been made that, as the ratio of EN to GJ, so the ratio of HF to some other line, that line is less than line GM. So let it be GO. And it is clear just as in the analysis that rectangle EF by OG is equal to rectangle FJ by JO.

So let another line be drawn, such as PX. Now it will be necessary to compare the ratio of EN to JG, that is, FH to GO, with the ratio of EX to PG. And by alternation, it will be necessary to compare the ratio of EX to FH, that is, the ratio of EP to PF, with the ratio of PG to GO. So by separation, it will be necessary to compare the ratio of EF to FP with the ratio of PO to OG. And it will be necessary to compare rectangle EF by GO with rectangle FP by PO. However rectangle EF by GO is equal to rectangle FJ to JO. So it will be necessary to compare rectangle FJ by JO with rectangle FP by PO.

And for that purpose it will be necessary to compare rectangle FK by KO with rectangle FJ by JO. However rectangle FJ by JO is equal to rectangle EF by OG. So it will be necessary to compare rectangle FK by KO with rectangle EF by GO. And the relation is this: that rectangle FK by KO is greater than rectangle EF by OG, as will be made clear, because rectangle EF by GM is greater than rectangle EF by GO, while rectangle EF by GM is equal to rectangle FK by KM, and rectangle EF by GO is equal to rectangle FJ by JO, so rectangle FK by KM is greater than rectangle FJ by JO. So rectangle FK by KO is much greater than rectangle FJ by JO.

And for that reason, rectangle FJ by JO is greater than rectangle FP by PO. So the ratio of EF to FP is greater than the ratio of PO to OG. So by composition, the ratio of EP to PF, that is, the ratio of EX to FH, will be greater than the ratio of PG to GO. And by alternation, the ratio of EX to PG is greater than the ratio of FH to GO, that is, the ratio of EN to JG. And for that reason, line NJ will cut off a ratio less than the ratio that line PX cuts off. And likewise, the lines that are nearer to line KL will cut off ratios less than the ratios that the more remote lines cut off.

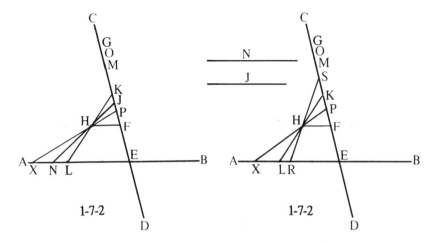

1-7-2   1-7-2

And the problem will be synthesized thus: Let the parallel line remain the same. And let there be taken a mean proportional between the two lines EG & EF, namely, EK. And let KL be joined. Now, line KL cuts off a ratio of EL to KG that is less than the ratios that all the lines drawn from point H cut off from the two lines GF, EA. So the given ratio is either the same as the ratio of EL to KG, or less than it, or greater than it.

So if it is the same as it, then line KL will make the problem work. And it will be clear that it alone does so, because the lines that are drawn from H alongside it cut off ratios greater than the one KL cuts off.

And if the ratio is given smaller, the problem will not obtain, because it is given less than the least.

Then let it be greater than the ratio of EL to KG, such as N to J. And let line KM be made equal to line FK. So rectangle EF by MG is equal to rectangle FK by KM. And the ratio of EL to KG is the same as the ratio of FH to GM. And the ratio of N to J is greater than the ratio of EL to KG, that is, the ratio of FH to GM. So let it be made that, as the ratio of N to J, so the ratio of FH to another line that is less than GM, and let it be GO. And since rectangle EF by GM is equal to rectangle FK by KM, rectangle EF by GO is less than rectangle FK by KO. So it is possible to apply to line OF a rectangle equal to rectangle EF by GO and defective by a square, in two ways, namely, on each of the two sides of point K. And the points of the <application> will be the two points S, P. And let SH, PH both be joined and produced. Then I say that each one of the two lines RS, PX will make the problem work. That is, that the ratio of EX to PG is like the ratio of N to J, and the ratio of ER to SG is like the ratio of N to J.

For, since rectangle EF by GO is equal to rectangle FP by PO, the ratio of EF to FP is like the ratio of PO to OG. And by composition and alternation, the ratio of EX to PG will be like the ratio of FH to GO, that is, the ratio of N to J. So line PX will make the problem work. And likewise, it is clear that line RS will make it work. So each one of the two lines RS, PX will make the problem work.

And it is clear that these two alone will do so, because a line nearer HK will cut off a ratio less than the ratio that the one farther away will cut off.

And we will find the ratio thus: Since the least ratio is the ratio of EL to KG, that is, the ratio of FH to GM, and GM is the excess of GE & EF together over ME & EF together, and ME & EF together are twice KE, because line MK is equal to line KF, and since the double of KE equaled in square four times rectangle GE by FE, because they are proportional: so for that reason, the least ratio is the ratio of FH to the excess by which GE & EF together exceed the line that equals in square four times rectangle GE by EF.

Case 3. Let the parallel line be the same. And let a line KL be drawn in the third way, cutting off from the two lines GE, EB a ratio of LE to KG the same as a given ratio. And let the ratio of FH to GM be made like the ratio of LE to KG. However line FH is a recipient. So line GM is a recipient in position and in magnitude. But point G is a recipient. So point M also is a recipient. So line FM. And since the ratio of LE to KG is like the ratio of FH to GM, then, by alternation, the ratio of LE to HF, that is, the ratio of EK to KF, will be like the ratio of KG to GM. And by composition, the ratio of EF to FK will be like the ratio of KM to MG. So rectangle FE by MG is equal to

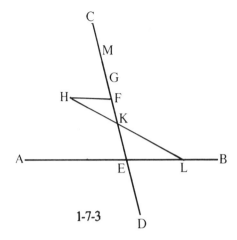

1-7-3

rectangle MK by KF. However rectangle EF by GM is a recipient, because each of the two sides is a recipient. So rectangle MK by KF is a recipient. And it has been applied to the recipient line MF with the excess of a square. So point K is a recipient. However point H also is a recipient. So line KL is positioned.

And this problem will be synthesized thus: Let the parallel line be as before. And the given ratio will be the ratio of N to J. And let the ratio of FH to GM be made like the ratio of N to J. And let there be applied to line MF a rectangle equal to rectangle FE by MG with the excess of a square, namely, rectangle MK by KF. And let HK be joined and produced in a straight line. Then I say that line HL will make the problem work. That is, that the ratio of N to J is like the ratio of LE to KG.

For, since rectangle EF by MG is equal to rectangle MK by KF, the ratio of EF to FK will be like the ratio of KM to MG. And by separation, the ratio of EK to KF, that is, the ratio of LE to HF, will be like the ratio of KG to GM. And by alternation, the ratio of LE to KG is like the ratio of FH to GM, that is, the ratio of N to J. So line HL will make the problem work.

And I say that it alone does that. For if it were possible, let another line be drawn, such as HP. So, if line HP cuts off a

ratio which is the ratio of N to J, then the ratio of LE to KG
will be like the ratio of PE to OG. But that is something which
is not possible. For the antecedent is less than the antecedent
while the consequent is greater than the consequent. And from
that it is clear that line HP cuts off a ratio less than the ratio
that line HL cuts off.

Case 4. Let the parallel line be as before. And let a line HK be
drawn in the fourth way, cutting off from the two lines EA, GD
a ratio of EL to KG the same as a given ratio. And let the ratio
of FH to GM be like the ratio of EL to KG. So FH is a
recipient. So line GM is a recipient in position and in
magnitude. But point G is a recipient. So point M is a recipient.
But point F is a recipient. So line FM is a recipient. And since
the ratio of FH to GM is like the ratio of EL to KG, then, by
alternation, the ratio of FH to EL, that is, the ratio of FK to
KE, is like the ratio of MG to GK. And by conversion, the ratio
of GM to MK will be like the ratio of KF to FE. So rectangle
GM by FE is equal to rectangle FK by KM. However rectangle
FE by GM is a recipient. So rectangle FK by KM is a recipient.
And it has been applied to the recipient line FM deficient by a
square. So point K is recipient. But point H is recipient. So line
HL is positioned.

And since in the synthesis the ratio of FH to GM must be
like the given ratio, and there must be applied to FM a
rectangle, like rectangle FK by KM, that is equal to rectangle
GM by FE and deficient by a square, the application is not
always possible, because what has been stated about the ratio
restricts its domain. So for that reason it is not always possible
for the problem to be synthesized in every way.

But let it obtain first in one way, namely, if point K is at
the midpoint of FM. So there will be a problem like this one:
How will we find a ratio, such that, when the ratio of FH to
GM has been made the same as it, and FM has been cut in half
at point K, the rectangle GM by FE will be equal to the
rectangle FK by KM? So let that be the case, and let us take a
point on line DG like M and cut line FM in half at point K,

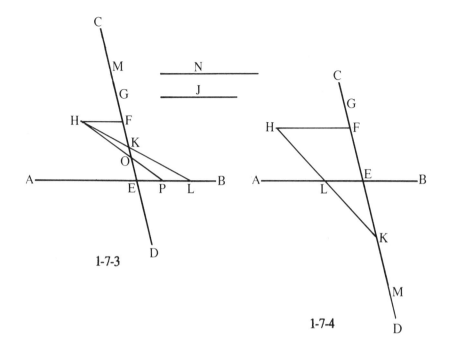

1-7-3

1-7-4

while the rectangle GM by FE is equal to rectangle FK by KM. Now, since rectangle GM by FE is equal to rectangle FK by KM, the ratio of GM to KM will be the same as the ratio of KF to FE. And by separation, the ratio of GK to KM will be like the ratio of KE to EF. However line MK is equal to line FK. So the ratio of GK to KF is like the ratio of KE to EF. So, by alternation, the ratio of GK to KE will become like the ratio of KF to EF. And by separating, the ratio of GE to EK will be like the ratio of KE to EF. So line EK is a mean proportional between the two lines GE, EF. And each one of the two lines GE, EF is a recipient. So line EK is a recipient in position and in magnitude. But point F is a recipient. So line FK is a recipient. And it is equal to line KM. So line KM is a recipient in position and in magnitude. But point K is a recipient. So point M is a recipient. And what is sought for is point M.

And this analysis will be synthesized thus: With the parallel line as before, let there be taken a mean proportional between the two lines GE & EF, namely, line EK. And let line KM be made equal to line FK. And let KH be connected. Then I say that rectangle GM by EF is equal to rectangle FK by KM, and that the ratio of EL to KG is the same as the ratio of FH to GM.

For, since the ratio of GE to EK is the same as the ratio of EK to EF, and the whole to the whole is like one of the single ratios, so the ratio of GK to KF is like the ratio of KE to EF. However line KF is equal to line KM. So the ratio of line GK to line KM is like the ratio of KE to EF. And by composition, the ratio of GM to MK will be like the ratio of KF to FE. So rectangle GM by EF is equal to rectangle FK by KM. And also, since the ratio of GM to MK is like the ratio of KF to FE, then, by conversion, the ratio of MG to KG is like the ratio of FK to KE, that is, the ratio of FH to EL. And by alternation, the ratio of FH to GM will be like the ratio of EL to GK. So before synthesizing, it will be necessary to take a mean proportional between the two lines GE & EF, namely, line EK, and then to join HK.

Then, let us inquire whether the line HK cuts off a ratio of EL to GK less than, or greater than, those which all the lines drawn from point H cut off from the two lines GD, EA. And we will find the problem thus: With the parallel line being as before, let there be taken a mean proportional between the two lines GE & EF, namely, line EK. And let HK be connected. It will be necessary to find out whether line HK cuts off a ratio of EL to GK that is greater or less than all the lines that are drawn cutting off from GD, EA. Now let line KM be made equal to line FK. So rectangle GM by EF is equal to rectangle FK by KM. And the ratio of EL to GK is the same as the ratio of FH to GM.

Then let another line be drawn, such as HN. So it will be necessary to compare the ratio of EL to GK, that is, the ratio of FH to GM, with the ratio of EJ to GN. And by alternation, it will be necessary to compare the ratio of FH to EJ, that is, FN to NE, with the ratio of MG to GN. And by conversion, it will be necessary to compare the ratio of GM to MN with the ratio of NF to FE. And it will be necessary to compare rectangle GM

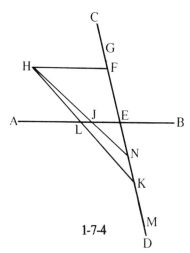

1-7-4

by EF with rectangle FN by NM. However rectangle FK by KM
is equal to rectangle GM by EF. So it will be necessary to
compare rectangle FK by KM and rectangle FN by NM. And
the relation is this: that rectangle FK by KM is greater than
rectangle FN by NM.

But rectangle FK by KM is equal to rectangle GM by FE.
So rectangle GM by FE is greater than rectangle FN by NM. So
the ratio of NF to FE is less than the ratio of GM to NM. And
by conversion, the ratio of FN to NE, that is, the ratio of FH to
JE, will be greater than the ratio of MG to GN. And by
alternation, the ratio of FH to GM, that is, EL to GK, is greater
than the ratio of EJ to GN. So line HK cuts off a ratio greater
than line HN. And likewise, it is clear that, among all those
drawn from point H cutting off the two lines GD & EA, line
HK will cut off a ratio of EL to KG that is greater than they
are.

And I say that the lines that are near line HK cut off ratios
greater than those that are distant from it. And since the ratio
of EL to KG, that is, FH to GM, is greater than the ratio of EJ
to GN, then, when we make the ratio of EJ to GN like the ratio
of FH to another line which is a greater one than GM, let it be
to GO. And likewise it will be clear that rectangle GO by FE is
equal to rectangle FN by NO.

Then let another line be drawn, such as HP. So it will be

necessary to compare the ratio of EJ to GN, that is, the ratio of FH to GO, with EX to GP. And by alternation, it will be necessary to compare the ratio of FH to EX with the ratio of OG to GP. However the ratio of FH to EX is like the ratio of FP to PE. So it will be necessary to compare the ratio of FP to PE and the ratio of OG to GP. And by conversion, it will be necessary to compare the ratio of PF to FE and the ratio of GO to OP. And it will be necessary to compare rectangle GO by EF and rectangle FP by PO. However rectangle GO by FE is equal to rectangle FN by NO. So it will be necessary to compare rectangle FN by NO and rectangle FP by PO.

And for that purpose it will be necessary to compare rectangle FK by KO and rectangle FN by NO. However rectangle FN by NO is equal to rectangle GO by FE. And so it will be necessary to compare rectangle FK by KO and rectangle GO by FE. And their relation is this: that rectangle FK by KO is greater than rectangle GO by FE, as we have explained, because rectangle OM by KF is greater than rectangle OM by EF, while rectangle MK by KF is equal to rectangle MG by FE, and so the whole rectangle FK by KO is greater than the whole rectangle GO by FE, that is, than rectangle FN by NO.

And for that reason, rectangle FN by NO is greater than rectangle FP by PO. So rectangle GO by FE is greater than rectangle FP by PO. So the ratio of FP to FE is less than the ratio of GO to PO. And by conversion, the ratio of FP to PE, that is, the ratio of FH to EX, is greater than the ratio of OG to GP. And by alternation, the ratio of FH to GO, that is, the ratio of EJ to GN, will be greater than the ratio of EX to GP. And for that reason, line HN will cut off a ratio that is greater than that which HP cuts off. So that the one near to line HK will cut off a ratio greater than the one distant from it.

This problem will be synthesized thus: With the parallel line as before, let there be taken a mean proportional between the two lines EG & EF, namely, line EK. And let HK be joined. So line HK will cut off a ratio of EL to GK that is greater than all that are drawn from point H and cut off from the two lines GD, EA. So the ratio being given in the synthesis is either the ratio of EL to GK, or else it is greater, or less.

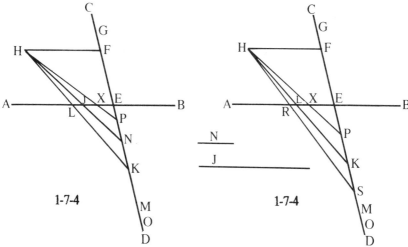

1-7-4   1-7-4

So, if it is the ratio of EL to KG, then line HK will make the problem work.

And if it is more than the ratio of EL to KG, then the problem will not be possible, because it is greater than the greatest.

So let the ratio of N to J be given less than the ratio of EL to KG. And let line KM be made equal to line FK. So rectangle GM by FE is equal to rectangle FK by KM, and the ratio of EL to KG is the same as the ratio of FH to GM. And since the ratio of N to J is less than the ratio of EL to KG, that is, than FH to GM, when we have made the ratio of FH to another line greater than GM be an instance of the ratio of N to J, then let FH have this ratio to GO. And since rectangle FK by OM is greater than rectangle FE by OM, while rectangle FK by KM is equal to rectangle MG by FE, the whole rectangle FK by KO is greater than the whole rectangle GO by FE. So it is possible to apply to line FO a rectangle equal to GO by FE and deficient by a square, in two ways, namely, on each side of point K. And the two points of the <application> will be the two points P & S. And let HP, HS be joined. Then I say that each one of the two lines HP, HS will make the problem work. That is, that the ratio of N to J is like the ratio of EX to GP, and that the ratio of N to J is like the ratio of ER to GS.

For, since rectangle FP by PO is equal to rectangle GO by FE, the ratio of GO to PO will be like the ratio of FP to FE.

And by conversion, the ratio of OG to GP will be like the ratio of FP to PE, that is, like the ratio of FH to EX. And by reverse alternation, the ratio of FH to GO will be like the ratio of EX to GP. But the ratio of FH to GO is like the ratio of N to J. So the ratio of N to J is like the ratio of EX to GP. And likewise it is clear that the ratio of ER to GS is like the ratio of N to J. So each one of HP, HS will make the problem work. And it is clear that they alone do so, because the ones close to line HK cut off ratios greater than the ones distant from it.

And we find the ratio thus: Since the greatest ratio is the ratio of EL to GK, that is, the ratio of FH to GM, or to GK & KF together, because MK is equal to FK, and the two lines GK & KF together are equal to the two lines GE & EF and the line EK taken twice, while twice EK equals in square four times rectangle GE by EF: so for that reason the ratio of FH to GM is either like the ratio of FH to the composite of GE & EF together with the line equaling in square four times the rectangle GE by EF, or the given ratio is less than that ratio.

Since we have made clear how we will synthesize the problem according to all the cases into which it is divided, next we will explain how many times it is possible for it to be synthesized for each of its domains. With the parallel line being as before, let there be taken a mean proportional line between the two lines GE, EF. And that will be either of the two lines EK, EN. And let HK, HN be joined. And let line KM be made equal to line FK, and line NO equal to line FN. So the ratio of EL to KG, that is, the ratio of FH to GM, is the least ratio according to the second case. And the ratio of EJ to NG, [that is, the ratio of FH to GO,] is the greatest according to the fourth case. And it will be clear that, according to the second case, the ratio of FH to GM will be greater than [the ratio of FH] to GO. So the given ratio is either like the ratio of FH to GM, or greater than the ratio of FH to GM, or less than the ratio of FH to GM but greater than the ratio of FH to GO, or like the ratio of FH to GO, or less than the ratio of FH to GO. If it is the ratio of FH to GM, then the problem will obtain

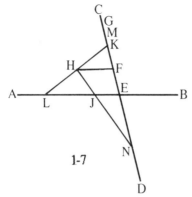

1-7

in three ways: according to the first case and the third, and once in the second case. But it will not obtain according to the fourth, because the ratio of FH to GM is greater than the ratio of FH to GO.

If it is greater than the ratio of FH to GM, then the problem will obtain in four ways: according to the first and third cases, and twice in the second case. But it will not obtain according to the fourth, because the given ratio is greater than FH to GM, and FH to GM is greater than the ratio of the fourth case.

If it be given less than the ratio of FH to GM, but greater than the ratio of FH to GO, then it will obtain in two ways: according to the first and third cases. But it will not obtain according to the second case nor according to the fourth, because the given ratio is less than the least in the second case, and greater than the greatest in the fourth case.

If it be the same as the ratio of FH to GO, then the problem will obtain in three ways: according to the first and third cases, and once in the fourth. But it will not obtain according to the second case, because the given ratio is less than the least.

And if it be given less than the ratio of FH to GO, then the problem will obtain in four ways: according to the first and third cases, and twice in the fourth case, but it will not obtain according to the second case, because it was given less than the least.

So we have clarified things according to the variety of ratios.

BOOK TWO

Now let there be two lines recipient in position, such as AB & DE, which cut one another at M. On line AB let a point C be positioned, and on line DE a point G.

Locus 1. And first let the recipient point H be within angle DMB. Then the lines drawn from point H cutting off a ratio the same as a given ratio fall in five ways. Either they cut off from the two lines CB, GE; or from lines CB, MG; or from lines GD, CA; or from lines GD, CM; or from lines DG, CB.

Case 1. So let a straight line HX be drawn in the first way, cutting off from the two lines CB, GE a ratio of CX to GP the same as a given ratio. And let CH be joined. Now, since point C is a recipient, and point H also, line HC will be positioned. But the line ED is positioned. So point F is a recipient. And let a line KL be drawn through point F parallel to line AB. Now, since a line KL has been drawn through a recipient point F parallel to a recipient line AB, and the points C, F, H are recipients, and also each of the two lines CH & HF, and so the ratio of CH to HF: then, since the ratio of CX to FR is a recipient, while the ratio of CX to GP is a recipient, so the ratio of FR to GP will be a recipient. So since the two lines KL, DE are recipients, with the terminal point on line KL at point F and the terminal point on line DE at point G, and the recipient point H is within angle DFL: then, if a line such as HR is drawn cutting off a ratio of FR to GP the same as a given ratio, the line HP is a recipient, as has been shown in the first case of the fourth figure of the first book. And that is what we intended to show.

And this problem will be synthesized thus: Let the ratio of N to O be given, and let it be made that, as the ratio of HC to HF, so the ratio of N to S. So since there are two recipient lines in a plane, namely, lines KL & DE, with the terminal point on

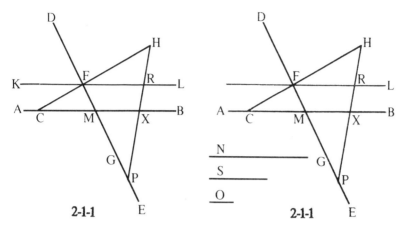

2-1-1   2-1-1

KL at point F and the terminal point on line DE at point G, and the recipient point H is within angle DFL, when the line HP has been drawn as specified in the first case of the fourth figure, cutting off a ratio of FR to GP like the ratio of S to O: then I say that line HP is the one that does what the problem requires.

For, since the ratio of HC to HF is like the ratio of CX to FR, while the ratio of HC to HF is like the ratio of N to S, so the ratio of CX to FR is like the ratio of N to S. However the ratio of FR to GP is like the ratio of S to O. So by way of equality, the ratio of CX to GP will be like the ratio of N to O. So line HP will do what the problem requires. And it is clear that it alone. But that is what we wanted to show.

Case 2. Then let HP be drawn according to the second case, cutting off from the two lines GM, CB a ratio of CX to GP the same as a given ratio. And let CH be joined. [Now, since point C is a recipient, and point H also, line HC will be positioned. But line DE also is positioned. So point F is positioned. And when a line such as KL is drawn through point F parallel to line AB, it will be positioned.] Now, since the points C, F, H are recipients, each of the two lines HC, HF is a recipient. So the ratio of HC to HF is a recipient. However the ratio of HC to HF is like CX to FR. So the ratio of CX to FR is a recipient.

But the ratio of CX to GP is a recipient. So the ratio of FR to GP is a recipient. So since there are two positioned lines KL, DE in a plane, with the terminal point on line KL at point F and the terminal point on line DE at point G, and the point H is within angle DFL, and a line HP has been drawn cutting off a ratio of FR to GP: then line HP will be a recipient, as we have explained in the second case of the fourth locus of the first book. And that is what we wanted to show.

In the synthesis of this it is necessary that the ratio being given be greater than the ratio of CM to MG. For, since line CX is greater than line CM, and line PG is less than GM, the ratio of CX to CM will be greater than the ratio of GP to GM. And by alternation, the ratio of CX to PG will be greater than the ratio of CM to MG. However the ratio of CX to PG is the same as the given ratio. So for that reason the ratio being given must be greater than the ratio of CM to MG.

And this problem will be synthesized thus: Let the rest of be as before, and let the given ratio be the ratio of N to O, which is greater than the ratio of CM to MG. Let CH and HM be joined. And let it be made that, as the ratio of CH to HF, so the ratio of N to S. So since the ratio of CH to HF is like the ratio of N to S, while the ratio of HC to HF is like the ratio of CM to FJ, the ratio of CM to FJ is like the ratio of N to S. But the ratio of N to O is greater than the ratio of CM to MG. So by way of equality, the ratio of S to O will be greater than the ratio of FJ to MG. So if we wanted to draw a line from point H according to the second case of the fourth locus, cutting off a ratio from the two lines KL, FG the same as the ratio of S to O, then, when the line was drawn, it would cut line GM. That is, point M would fall above, since it has become clear in the synthesis of the second case of the fourth locus that lines near point F cut off smaller ratios. So let line PH be drawn, cutting off a ratio of FR to PG the same as the ratio of S to O. Then I say that PH will do what the problem requires.

For, since the ratio of HC to HF is like the ratio of N to S, while the ratio of HC to HF [is like the ratio of CX to FR, so the ratio of CX to FR] is like the ratio of N to S. But the ratio

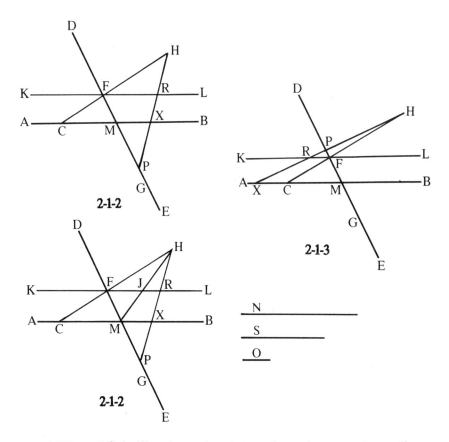

2-1-2

2-1-3

2-1-2

N

S

O

of FR to PG is like the ratio of S to O. So by way of equality, the ratio of CX to PG is like the ratio of N to O. So line HP will do what the problem requires. And it is clear that it alone. And that is what we wanted to show.

Case 3. Then let HP be drawn according to the third case, cutting off from the two lines CA, GD a ratio of CX to GP the same as a given ratio. And let CH be joined. Now, since point C is a recipient, and point H also is a recipient, line CH will be a recipient. However line DE is positioned. So point F is positioned. And let a line KL be drawn through point F parallel to line AB. So line KL is positioned. And since the line

KL and the points C, H, F are recipients, each of the two lines CH, HF will be recipients. And the ratio of CH to HF is like the ratio of CX to RF. So the ratio of XC to FR is a recipient. However CX to GP is a recipient. [So the ratio of FR to GP is a recipient.] So since there are two lines KL, DE recipient in a plane, with the terminal point on line KL at point F and the terminal point on line DE at point G, and the recipient point H is within angle DFL, and a line HP has been drawn cutting off the ratio of FR to PG, then line HP will be a recipient, as was made clear in the third case of the fourth locus of the first book. And that is what we wanted to show.

And this problem will be synthesized thus: Let the given ratio be the ratio of N to O, while the rest will be as before. And let it be made that, as the ratio of CH to HF, so the ratio of N to S. So since there are two recipient lines KL, DE in a plane, with the terminal point on line KL at point F and the terminal point on line DE at point G, and the recipient point H is within angle DFL: then let a line HR be drawn according to the third case of the fourth locus, cutting off a ratio of FR to GP the same as the ratio of S to O. And let the line HR be produced to X. Then I say that HX will do what the problem requires.

For, since the ratio of CH to HF is like the ratio of N to S, the ratio of CX to FR is like the ratio of N to S. But also, the ratio of FR to GP is like the ratio of S to O. So by way of equality, the ratio of CX to GP will be like the ratio of N to O. So line HX will do what the problem requires. And it is clear that it alone. And that is what we wanted to show.

Case 4. Then let HP be drawn according to the fourth case, cutting off from the two lines CM, GD a ratio of XC to GP the same as a given ratio. And let line CH be joined. Now, since point C is a recipient, and point H is a recipient, line CH is a recipient. But line DE is a recipient. So point F is a recipient. And let a line KL be drawn through point F parallel to line AB. So line KL is a recipient. And since the points C, H, F, are recipients, each of the two lines CH, HF will be a recipient. However the ratio of CH to HF is like the ratio of XC to FR.

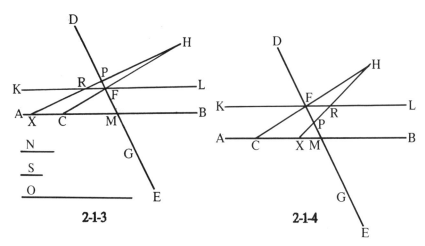

2-1-3   2-1-4

So the ratio of XC to FR is a recipient. But the ratio of XC to
GP is a recipient. So the ratio of FR to GP is a recipient. So
since there are two recipient lines KL, DE in a plane, with the
terminal point on KL at point F and the terminal point on DE
at point G, and the recipient point H is within angle DFL, and
a line HP has been drawn cutting off the ratio of FR to GP:
then line HP is a recipient, as we have explained in the second
case of the fourth locus of the first book.

Now, in the synthesis it will be necessary for the ratio
being given to be less than the ratio of CM to MG. For, since
line MC is longer than CX, and line MG is less than GP, the
ratio of MC to XC will be greater than the ratio of MG to GP.
And by alternation, the ratio of MC to MG will be greater than
the ratio of CX to GP. However the ratio of XC to GP is the
given ratio. So it will be necessary for the ratio being given in
the synthesis to be less than the ratio of CM to MG. And that is
what we wanted to show.

And this problem will be synthesized thus: Let the rest be
the same as it was, while the given ratio is the ratio of N to O,
which is less than the ratio of CM to MG. Let HM be joined,
and let it be made that, as the ratio of CH to HF, so the ratio
of N to S. So since the ratio of CH to HF is like the ratio of
CM to FJ, and the ratio of CH to HF is like the ratio of N to S,

the ratio of CM to FJ is like the ratio of N to S. But the ratio of CM to MG is greater than the ratio of N to O. So by way of equality, the ratio of FJ to MG is greater than the ratio of S to O. So when we have drawn from point H a line cutting off from the two lines GF, FL a ratio the same as the ratio of S to O, then that line will cut FM. For, since the two lines KL, DE are recipients, with the terminal point on line KL at point F and the terminal point on line DE at point G, and the recipient point is within angle DFL, and a line HM has been drawn according to the second case of the fourth figure, cutting off a ratio of FJ to MG which is greater than the ratio of S to O: then, if we wanted to draw another line that cuts off a ratio the same as the ratio of S to O, it would be a line falling in the position of HX. For lines near point F will always cut off ratios less than the ratios that lines that are distant from it cut off. So let the line be HR, and let it be produced in a straight line. Then I say that line HX will do what the problem requires.

For, since the ratio of CH to HF is like the ratio of CX to FR, and like the ratio of N to S, while the ratio of FR to GP is like the ratio of S to O, so by way of equality, the ratio of CX to GP will be like the ratio of N to O. So line HX will do what the problem requires. And that is what we wanted to show.

Case 5. And let PX be drawn according to the fifth case, cutting off from the two lines GD, CB a ratio of CX to GP the same as a given ratio. And let CH be joined. So since point C is recipient and point H is recipient, line CH is a recipient. But line DE is a recipient. So point F is a recipient. And through point F let a line be drawn parallel to line AB, according to which KL is a recipient. Then since points C, H, F are recipients, each of the two lines CH, HF will be a recipient. So the ratio of CH to HF is a recipient. However the ratio of CH to HF is like the ratio of CX to FR. So the ratio of CX to FR is a recipient. However the ratio of CX to GP likewise is a recipient. So the ratio of FR to GP is a recipient. So when there are two recipient lines, namely, lines KL & DE, and the terminal point on KL is at point F and the terminal point on line DE is at point G, and the recipient point H is within angle DFL, and there is a line, namely, line PX, which cuts off the

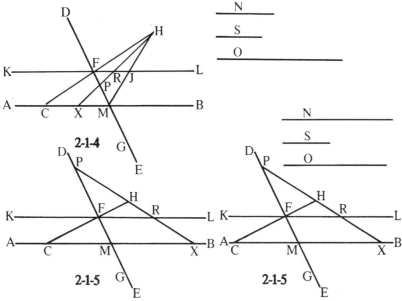

2-1-4

2-1-5

2-1-5

ratio of FR to GP: then line PR will be positioned, as has been shown in the fourth case of the fourth locus of the first book. And that is what we wanted to show.

The problem will be synthesized thus: Let the given ratio be the ratio of N to O, while things are the same as before. Let it be made that, as the ratio of CH to HF, so the ratio of N to S. So when there are two positioned lines in a plane, namely, the lines KL & DE, and the terminal point on line KL is at point F and the terminal point on line DE is at point G, and the recipient point H is within angle DFL, and the ratio is the ratio of S to O: then let line PR be drawn according to the fourth case of the fourth locus, cutting off a ratio of FR to GP the same as the ratio of S to O. And let PR be prolonged as line PRX. So I say that line PRX will do what the problem requires.

For, since the ratio of CH to HF is like the ratio of CN to FR, while the ratio of CH to HF is like the ratio of N to S, so the ratio of CX to FR is like the ratio of N to S. But the ratio of FR to PG is like the ratio of S to O. So, by way of equality, the ratio of CX to PG will be like the ratio of N to O. And that is what we wanted to show.

Let there be the two positioned straight lines. And let them both have terminal points on them, line AB at point C and line DE at point G. And let the recipient point H be within angle EMB.

Locus 2. And first let the line that is drawn through point H parallel to line AB fall on point G. Then the lines drawn through H will fall in four ways. Either they will cut off from the two lines CB, GE; or from the two lines CA, DG; or from the two lines CM, MG; or from the two lines CB, GD.

Case 1. So first let a line RHX be drawn according to the first case, cutting off from the two lines CB, GE a ratio of CX to GR the same as a given ratio. And let CH be joined. Now point H is a recipient. But point C is a recipient. And line GE is positioned. And line CH is. So point F is a recipient. And let a line KL be drawn through point F parallel to line AB. So line KL is positioned. And since each one of the two lines CH, HF is a recipient, the ratio of CH to HF is a recipient. However the ratio of CH to HF is like the ratio of CX to FP. So the ratio of CX to FP is a recipient. But the ratio of CX to GR is a recipient. So the ratio of FP to GR is a recipient. So, since the two lines KL, DE are recipient in position in a plane, while the terminal point on line KL is at point F and the terminal point on line DE is at point G, and the recipient point H is within angle EFL, and the line which passes through point H parallel to line AB falls first on point G, and a line XH has been drawn cutting off the ratio of FP to GR: then line XHR is positioned, as we have shown in the first case of the fifth locus of the first treatise. And that is what we wanted to show.

And the synthesis is as follows: Let the rest be as before, and let things be made according to the particular situation that was described in the analysis. And let the given ratio be the ratio of N to O, while it is made that, as the ratio of CH to HF, so the ratio of N to S. And since there are two recipient lines in a plane, namely, KL & DE, and the terminal point on line KL

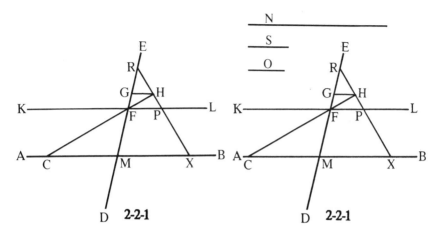

2-2-1   2-2-1

is at point F and the terminal point on line DE at point G, and the recipient point H is within angle EFL, and the ratio is the ratio of S to O: then let line PR be drawn according to the first case of the fifth locus, wherein the parallel line falls on point G, cutting off the ratio of FP to GR the same as S to O. And let line PHR be produced. Then I say that line PHR will do what the problem requires.

For, since the ratio of CH to HF is like the ratio of CX to FP, while the ratio of CH to HF is like the ratio of N to S, so the ratio of CX to FP is like the ratio of N to S. But the ratio of FP to GR is like the ratio of S to O. So, by way of equality, the ratio of CX to GR will be like the ratio of N to O. So line PHR will do what the problem requires. And that is what we wanted to show.

Case 2. Then let a line HP be drawn according to the second case, cutting off from the two lines CA, GD a ratio of CR to GP the same as a given ratio. And let CH be joined. So point F is a recipient. And let a line KFL be drawn through point F parallel to line AB. So line KFL is positioned. And since the ratio of CH to HF is a recipient, and the ratio of CH to HF is like the ratio of CR to FX, so the ratio of CR to FX is a recipient. However the ratio of CR to GP is a recipient. So the ratio of FX to GP is a recipient. So, since the two lines KL, DE are recipient, and the terminal point on line KL is at point F and the terminal point on line DE is at point G, and the recipient point H is within angle EFL, and the parallel line falls on point G, and a line HX has been drawn, cutting off a ratio from the two lines FK & GD: then the line HPX is positioned, as has been shown in the second case of the fifth locus of the first book. And this is what we wanted to show.

And the problem will be synthesized thus: Let the rest be as was described before, and let the given ratio be the ratio of N to O. And let it be made that, as the ratio of CH to HF, so the ratio of N to S. So, since there are two recipient lines in a plane, namely, lines KL & DE, and the place which is passed by on line KL is point F, which is one limit, and the limit on line DE is at point G, and the recipient point H is within angle EFL, and the parallel line falls on point G, and the ratio is the ratio of S to O: then let line HPX be drawn according to the second case of the fifth locus, cutting off a ratio of FX to GP the same as the ratio of S to O. And let it be produced to point R. Then I say that line HR will do what the problem requires.

For, since the ratio of CH to HF is like the ratio of CR to FX, while the ratio of CH to HF is like the ratio of N to S, so the ratio of CR to FX is like the ratio of N to S. And the ratio of FX to GP is like the ratio of S to O. So, by way of equality, the ratio of CR to GP is like the ratio of N to O. So the line HPR will do what the problem requires. And that is what we wanted to show.

Case 3. And let a line HP be drawn according to the third case, cutting off from the two lines CM, MG a ratio of CP to GX the same as a given ratio. And let CH be joined. So point F is a recipient. And through point F let a line KFL be drawn parallel to line AB. So line KFL is positioned. And since the ratio of CH to HF is like the ratio of PC to FR, the ratio of CP to FR is a recipient. So the ratio of FR to GX is a recipient. And since there are two positioned lines in a plane, namely, KL & DE, while the limit on KL is at point F and the limit on DE is at point G, and the recipient point H is within angle EFL, and the parallel line falls on point G, and a line HP has been drawn

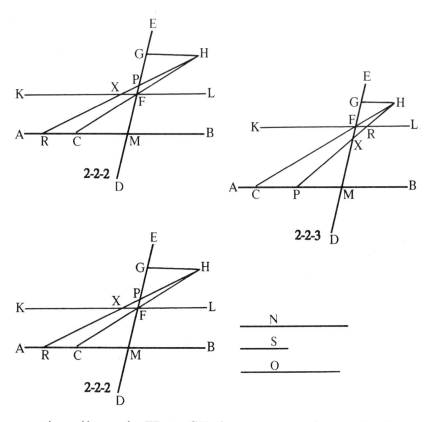

cutting off a ratio FR to GX the same as a given ratio: then line HXP is a recipient. And it will do what the problem requires, as has been shown in the third case of the fifth locus of the first book. And that is what we wanted to show.

And this will be found as follows: Let the rest be the same, and let line GF be either less than line FM, or larger than it. So first let it not be less than it. And let HM be joined. Then I say that line HM will cut off a ratio of CM to MG greater than all the ratios that lines drawn from point H cut off, while cutting the two lines CM, FM.

For let another line be drawn, such as line HXP. Then, since the relation of GF to FM is that it is not less, line HM either cuts off a ratio of FJ to GM which is greatest, or it is nearer to the line that cuts off the greatest ratio, and is greater than what line HXP cuts off, as has been shown in the third case of the fifth locus of the first book. So the ratio of line FJ to GM is greater than the ratio of FR to GX. And by alternation, the ratio of JF to FR is greater than the ratio of MG to GX. However the ratio of JF to FR is like the ratio of CM to CP, so the ratio of MC to CP is greater than the ratio of MG to GX. And by alternation, the ratio of MC to MG is greater than the ratio of CP to GX. So the line HM will cut off a ratio of CM to MG greater than all the ratios that the lines drawn through point H cut off, while cutting the two lines FM & CM, and that is what we wanted to show.

Now let line GF be less than line FM. And let line XF be made equal to line FG. And let HX be joined and produced to point P. And let MH be joined. Then I say that line XP cuts off a ratio CP to GX which is greater than all the ratios that the lines drawn from point H cut off, while cutting CM. And as for the lines drawn from point H that cut line PM, line HM cuts off a ratio of CM to MG, the least ratio.

So let another line be drawn that cuts CP, such as NO. Then, since line GF is equal to line FX, because it has been thus posited, line XP cuts off a greatest ratio of FR to GX, as we have shown in the third case of the fifth locus of the first book. So the ratio of FR to XG is greater than the ratio of FS to GO, as we have shown in the first treatise. And by alternation, the ratio of FR to FS is greater than the ratio of GX to GO. However the ratio of FR to FS is like the ratio of PC to CN. So the ratio of PC to CN is greater than the ratio of XG to OG. And by alternation, it is also thus. And if another line that cuts CP is drawn from point H, it will be clear that the ratio of CP to GX is greater than the ratio it cuts off.

And I say that line HM will cut off a ratio of CM to MG less than all the ratios that the lines drawn through point H cut off, while cutting off from the two lines PM, XM. So let another line that cuts PM be drawn, such as HTV. So, since HXP cuts off a greatest ratio, and line TUV is nearer to it than line HM, the ratio of TF to UG is greater than the ratio of JF to MG.

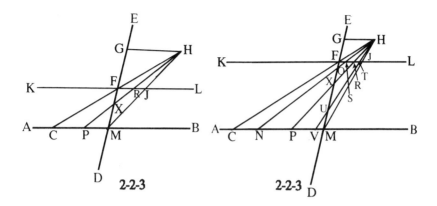

2-2-3   2-2-3

And by alternation, the ratio of TF to FJ is greater than the ratio of UG to GM. But the ratio of TF to FJ is like the ratio of VC to CM. So the ratio of VC to CM is greater than the ratio of UG to MG. And by alternation, the ratio of CV to UG is greater than the ratio of CM to MG. So the ratio of CM to MG is less than the ratio of CV to UG. So line HX will cut off a ratio of PC to XG greater than all the ratios that lines drawn from point H cut off, while cutting CM. And line HM will cut off a ratio of CM to MG less than all the lines that cut PM. And that is what we wanted to show.

And the synthesis of this problem is thus: Let the rest be the same, and let line GF be either less than line FM, or greater. So first let line GF not be less than line FM. And let HM be joined. So line HM will cut off a ratio of CM to MG greater than all the ratios that the lines drawn through point H cut off, while cutting line CM.

So, if the ratio being given in the synthesis is <the same as> the ratio of CM to MG, then line CM alone will do what the problem requires.

And if it is greater than the ratio of CM to MG, the problem will not be synthesized, because line CM will cut off the ratio of CM to MG, which is greatest.

And if it is less than the ratio of CM to MG, the problem will obtain in one way. For, let the given ratio be the ratio of N to O, which is less than the ratio of CM to MG. And let it be made that, as the ratio of CH to HF, so the ratio of N to S. And let a line KFL be drawn through point F parallel to line AB. So, since the ratio of CH to HF is like the ratio of CM to FJ, while the ratio of CH to HF is like the ratio of N to S, the ratio of CM to FJ is like the ratio of N to S. And inversely, the ratio of JF to CM is like the ratio of S to N. But the ratio of CM to MG is greater than the ratio of N to O. So, by way of equality, the ratio of FJ to MG will be greater than the ratio of S to O. So, since the two lines KL, DE are positioned, and the point of termination on line KL is at point F, while the point which is passed by on DE is at point G, and the parallel line falls on point G, and line FG is not less than line FM: then line HM cuts off a greatest ratio, namely, the ratio of FJ to MG, as has been shown in the third case of the fifth locus of the first book. And the ratio of S to O is less than the greatest ratio. So, if according to the third case of the fifth locus, we wanted to draw from point H a line which cuts off from the two lines FL, GM a ratio like the ratio of S to O, in two ways, then one line will cut [FM] and the other will pass by. Let the one be drawn. So line HRP will cut off a ratio of RF to GX the same as the ratio of S to O. So I say that line HNX will do what the problem requires.

For, since the ratio of CH to HF is like the ratio of PC to FR, and the ratio of CH to HF is like the ratio of N to S, so the ratio of PC to FR is like the ratio of N to S. But the ratio of RF to GX is like the ratio of S to O. So, by way of equality, the ratio of CP to XG will be like the ratio of N to O. So line HP will do what the problem requires. And that is what we wanted to show.

Now let line GF be less than line FM. And let line FX be made equal to line GF. And let HX be joined, and produced in a straight line to point P. And let HM be joined. So line HXP will cut off a ratio of PC to XG, which is greater than all the ratios that the lines drawn from point H cut off, while cutting

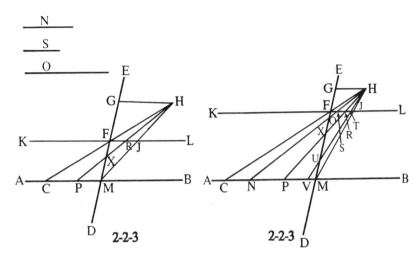

2-2-3

2-2-3

line CM. And line HM will cut off the ratio of CM to MG, less than all the ratios that the lines drawn from point H cut off, while cutting PM.

For, since line PXH, according to the third case of the fifth locus, cuts off the ratio of FR to XG, which is greatest, then the ratio of FR to XG is greater than the ratio of JF to MG. And by alternation, it will be so. But the ratio of RF to FJ is like the ratio of CP to CM. And so, by alternation, the ratio of PC to GX is greater than the ratio of CM to MG. So line HP cuts off a ratio of CP to GX greater than all the ratios that the lines which cut CM cut off.

And line HM will cut off a ratio of CM to MG less than all the ratios that the lines cutting PM cut off. For let line HTU be drawn. So, since line HTU is nearer to the line that cuts off the greatest ratio than line HM is, then line HTU will cut off a ratio greater than what HM cuts off. So the ratio of TF to GU will be greater than the ratio of JF to MG. And by alternation, it will be so. And the ratio of line TF to FJ is like the ratio of VC to CM. And, according to alternation, the ratio of VC to GU is greater than the ratio of CM to MG. So line HM cuts off the least of the ratios.

And for that reason, if in the synthesis a ratio known to be the same as the ratio CP to XG is given, then line HP alone will do what the problem requires. And if the ratio is given greater than the ratio of CP to GX, the problem will not be synthesized, because the ratio is too great. And if it is less than the ratio of CP to XG and greater than the ratio of CM to MG, then the problem will be synthesized in two ways, because it will be drawn on both sides of HXP while doing what the problem requires. And if the ratio is not greater than the ratio of CM to MG, the problem will obtain in one way, in the interval between C & P.

So first let the given ratio be the ratio of N to O, which is less than the ratio of CP to GX and greater than the ratio of CM to MG. And let it be made that, as the ratio of CH to HF, so the ratio of N to S. So the ratio of PC to RF will be like the ratio of N to S. And according to the reverse of that, the ratio of RF to PC will be like the ratio of S to N. But the ratio of CP to XG is greater than the ratio of N to O. And by way of equality, the ratio of RF to XG is greater than the ratio of S to O.

And also, since the ratio of CH to HF is like the ratio of CM to JF, so the ratio of CM to JF is like the ratio of N to S. So, by the reverse of that, the ratio of JF to MC is like the ratio of S to N. But the ratio of CM to MG is less than the ratio of N to O. So, by way of equality, the ratio of JF to MG will be less than the ratio of S to O. And it has become clear that the ratio of RF to GX is greater than the ratio of S to O. And as for line HX, it cuts off a greatest ratio according to what has been stated in the third case of the fifth locus. So, if we wanted to draw lines from point H according to the third case of the fifth locus, cutting off from the two lines FJ, GM lines whose ratio to one another is like the ratio of S to O, two lines will be drawn, one on each side of line HP. So let a line HON be drawn which cuts line FX. Then I say that the other line will cut XM.

For, since what is nearer the line PH will cut off a ratio greater than the lines that are farther from it, and the ratio of S to O is greater than the ratio of JF to MG, then the line which is drawn from point H and cuts off the ratio is nearer to the line HP than HM is. So let it be line HTU. And let it be produced in a straight line to V. Then I say that HV will do

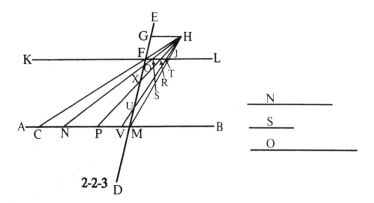

what the problem requires.

For, since the ratio of CH to HF is like the ratio of CV to FT, the ratio of N to S is like the ratio of CV to FT. But the ratio of S to O is like the ratio of FT to UG. So, by way of equality, the ratio of N to O will be like the ratio of VC to UG. So line HUV will do what the problem requires. And it will be clear that only it and the other line, that is, HP, will do what the problem requires.

Let everything be as it was, and let the ratio of N to O be not greater than the ratio of CM to MG. And let it be made that, as the ratio of CH to HF, so the ratio of N to S. If we inquire as we did before, S to O is not greater than the ratio of FJ to MG, which is not greater than the ratio of FR to GX. So, since the ratio of S to O is less than the ratio of RF to GX, that is, than the greatest, as was said in the third case of the fifth locus, let two lines be drawn, on the two sides of line HXP, cutting off a ratio the same as S to O. So the one will cut in the interval between the two lines PH & CH, and the other will <pass by>.

For, since it is always the line nearer to line HRX which cuts off a greater ratio, and the ratio of S to O may be compared as not being greater than the ratio of FJ to MG, either it is the same as it or it is less than it. So, if it is the same as it, then the <sought> line is line HM. And if it is less than the ratio of FJ to GM, then the line falls outside of line CPM, <and it is one> that cuts off but does not do what the problem requires. For it was <necessary that it be one> that cuts CM. And as for the line which cuts between the two lines CH & HP, it will do what the problem requires. And that is what we wanted to show.

Case 4. And let a line HP be drawn according to the fourth case, cutting off from the two lines CB, GD a ratio of CX to PG the same as a given ratio. And let CH be joined. So point F is recipient. And through point F let a line be drawn parallel to line AB. So line KL is positioned. And since the ratio of CH to HF is a recipient, while the ratio of CH to HF is like the ratio of CX to FR, so the ratio of CX to FR is a recipient. [So the ratio of PG to FR is a recipient.] So since the two lines KL, DE are positioned, and the point which is passed by on line KL is at F, and the point which is passed by on line DE at point G, and the recipient point H is within angle EFL, and the parallel line falls on the recipient point G, and a line HRP has been drawn, cutting off the ratio of PG to FR: so line HP is positioned, as has been shown in the third case of the fifth locus of the first book. And that is what we wanted to show.

And we will find this problem thus: Line FG is either greater than line FM or it is less than it. So first let line GF not be greater than line FM. And let HM be joined. Then I say that the ratio of line CM to line MG is greater than all the ratios that all the lines drawn from point H cut off, while cutting DM.

For let another line be drawn, such as as HXP. Then, since line GF is not longer than line FM, HM either will cut off a ratio of FJ to MG, which is greatest, or it will be nearer to the line that cuts off a greatest ratio than HP is. So the ratio of JF to MG is greater than the ratio of FR to GP. And by alternation, the ratio of JF to FR is greater than the ratio of MG to GP. However the ratio of JF to FR is like the ratio of CM to CX. So the ratio of MC to CX is greater than the ratio of MG to GP. And by alternation, the ratio of MC to MG is greater than the ratio of CX to GP. So line HM will cut off a

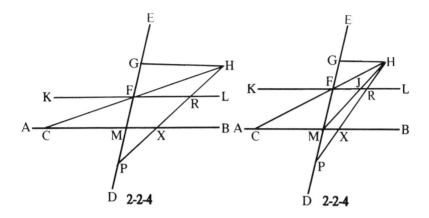

D 2-2-4   D 2-2-4

ratio of CM to MG which is greater than all the ratios that the lines drawn from point H cut off, while cutting MD. And that is what we wanted to show.

And let things be as they were, except let line GF be longer than line FM. And let line FP be made equal to line GF. And let HP be joined. Then I say that line HP will cut off a ratio of CX to GP greater than all the ratios that the lines drawn from point H cut off, while cutting line MD.

For let two lines HN, HV be drawn on either side of HP. Then, since line GF is equal to line FP, line HP will cut off a ratio of CX to PG, which is a greatest ratio according to the example of the third case of the fifth locus. For, since two lines KL, DE are positioned, and the terminal point on line KL is at point F, and on line DE at point G, and the parallel drawn through the recipient point H has fallen on point G, and line FP is equal to line GF: so the ratio of RF to PG is greater than the ratio of SF to GN. And by alternation, the ratio of RF to FS is greater than the ratio of PG to NG. However the ratio of RF to FS is like the ratio of CX to CO. So the ratio of CX to CO is greater than the ratio of PG to NG. [And by alternation, the ratio of CX to PG will be greater than the ratio of CO to NG.] And for <a similar reason> it will become clear that line HV will cut off a ratio less than the ratio that line HP cuts off. So line HP will cut off a ratio of <CX> to GP which is greater than all the ratios that lines drawn from point H cut off, while cutting MD.

And I say that line HM will cut off a ratio of CM to MG which is less than the ratios that the drawn lines cut off, while cutting PM only. For, since line HV is nearer to the line that cuts off the <greatest> ratio, that is, line HP, than line HM is, the ratio of TF to VG will be greater than the ratio of JF to MG. And by alternation, the ratio of TF to FJ is greater than the ratio of VG to GM. However the ratio of TF to FJ is like the ratio of UC to CM. So the ratio of UC to CM is greater than the ratio of VG to GM. And by alternation, the ratio of UC to VG will be greater than the ratio of MC to MG. So line HM will cut off a ratio of CM to MG which is less than all the ratios which the lines drawn from point H cut off, while cutting line PM alone. And that is what we wanted to show.

And this problem will be synthesized thus: Let the rest be as before. Then line GF will either be longer than line FM or it will be shorter than it. So first let line GF be not longer than line FM. And let HM be joined. Then line HM will cut off a ratio of CM to MG, which is greater than all the ratios that the lines drawn from point H cut off, while cutting line DG.

Then, if the given ratio is the same as the ratio of CM to MG, line HM alone will do what the problem requires, because it is the one that cuts off the greatest ratio.

And if it is greater, then the problem will not be synthesized, because it will be given greater than the greatest.

But if it is given less, then the problem will obtain in one way. For let the given ratio be the ratio of N to O, which is less than the ratio of CM to MG. And let it be made that, as the ratio of CH to HF, so the ratio of N to S. Then, since the ratio of CH to HF is like the ratio of CM to FJ, and like the ratio of N to S, so the ratio of CM to FJ is like the ratio of N to S. However the ratio of CM to MG is greater than the ratio of N to O. So, by way of equality, the ratio of S to O will be less than the ratio of FJ to MG. And the ratio of FJ to MG is the greatest ratio that existed in the third case of the fifth locus. So two lines may be drawn from point H, one on each side of line HM, that will cut the two lines MD, MG with a ratio the same

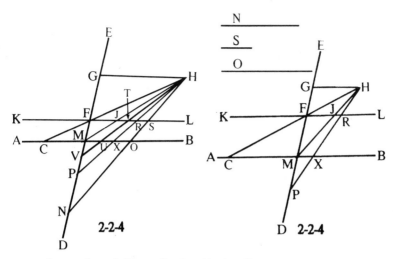

2-2-4

D 2-2-4

as the ratio of S to O. So, if the line cutting MD is drawn, namely, the line HP, which cuts off the ratio of RF to GP the same as the ratio of S to O, it will do what the problem requires.

For, since the ratio of CH to HF is like the ratio of N to S, while the ratio of CH to HF is like the ratio of CX to FR, and the same as the ratio of N to S, and the ratio of RF to PG is like the ratio of S to O, so, by way of equality, the ratio of CX to PG is like the ratio of N to O. So line HP will do what the problem requires. And as for the other line, since it cuts line CM, it will not do what the problem requires. And for that reason the problem will not exist except in one way. And that is what we wanted to show.

Let things be the same as they were, except that line GF will be longer than line FM, and line FP will be made equal to it. And let the two lines HM, HP be joined. So, since line GF is equal to line FP, line HP will cut off a ratio of CX to PG which is greater than all the ratios that the lines cutting MD cut off. And line MH will cut off a ratio of CM to MG which is less than all the ratios that the lines that cut line PM cut off.

So if the given ratio is the same as the ratio of CX to PG, then the line HP alone will do what the problem requires. And if it is given greater than the ratio of CX to PG, then the problem will not be synthesized, because the ratio has been given greater than the greatest. And if it was given less than the ratio of CX to PG but greater than the ratio of CM to MG, then the problem will be synthesized in two ways, because two lines will be drawn, on both sides of PH, which will cut MD and do what the problem requires. And if the ratio is not greater than the ratio of CM to MG, then the problem will exist only in one way.

So let the given ratio be the ratio of N to O, which is less than the ratio of CX to PG and greater than the ratio of CM to MG. And let the ratio of CH to HF be made like the ratio of N to S. Then, since the ratio of CH to HF is like the ratio of CX to FR, and like the ratio of N to S, so the ratio of CX to FR is like the ratio of N to S. But the ratio of CX to PG is greater than the ratio of N to O. So, by way of equality, the ratio of RF to GP will be greater than the ratio of S to O.

And also, since the ratio of CH to HF is like the ratio of N to S, so the ratio of N to S is like the ratio of CM to FJ. However the ratio of CM to MG is less than the ratio of N to O. So, by way of equality, the ratio of FJ to MG will be less than the ratio of S to O.

And since the ratio of S to O is less than the ratio of RF to PG and greater than the ratio of FJ to MG, while the ratio of RF to PG is the greatest ratio that obtains in the third case of the fifth locus, so two lines may be drawn through point H, on both sides of PH, which cut off line MD with a ratio the same as the ratio of S to O. So, if the two lines VH, NH are both drawn, they will both do what the problem requires.

For, since the ratio of CH to HF is like the ratio of N to S, and the ratio of CH to HF is like the ratio of CO to FS, so the ratio of CO to FS is like the ratio of N to S. However the ratio of FS to NG is like the ratio of S to O. So line NH will do what the problem requires. And similarly, too, it will become clear that line VH will do what the problem requires.

And how will line HV not fall outside of line HM? This is clear as follows: Since the ratio of S to O is less than RF to PG, which is the greatest ratio that obtained in the third case of the fifth locus, and greater than that which line HM cuts off,

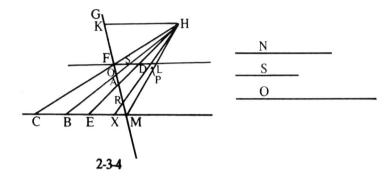

2-3-4

namely, the ratio of JF to MG, and since that which is nearer to
the line which cuts off the greatest ratio cuts off a ratio greater
than the ratio which the more distant line cuts off: so the line
that is drawn and cuts off a ratio like S to O will cut PM. And
that is what we wanted to show.

Let things be the same, and let the given ratio be not
greater than the ratio of CM to MG, but let it be the same as
the ratio of N to O. And let things be as they were. Then it will
be clear that the ratio of S to O is not greater than the ratio of
FJ to GM. However it is less than the ratio of FR to GP, which
is the greatest ratio in the third case of the fifth locus. So two
lines may be drawn, on either side of line HP, which both cut
off a ratio the same as the ratio of S to O. And one of them
will cut off DM, and will do what the problem requires, as was
previously explained. But the other one will not do what the
problem requires, because it will not cut line PM, but line MF.
This is because the lines near line HP cut off ratios greater
than the ratios which the lines remote from it cut off, and the
ratio of S to O is less than the ratio of FJ to MG, <and the
other line is distant from line HP by a greater interval than
line HM is.> So <because of its distance> it will cut MF. And
that is what we wanted to show.

Locus 3. And also let the line which is drawn through point H parallel to line AC fall below point G, that is, between it and point M, as line HK. Then it will also be clear that the lines drawn from point H fall in five positions.

Case 1. So let a line AB be drawn according to the first case, cutting off a recipient ratio of CA to BG. Let CH be joined, and let a line SFO be drawn through point F parallel to line CA. Then, since the ratio of CH to HF is a recipient, and thus the ratio of CA to OF is a recipient, so the ratio of OF to GB is a recipient. So, since there are two positioned lines, namely, SO & MB, and the place which is passed by on line MB is at point G, and the place which is passed by on line SO is at point F, and the recipient point H is within angle OFB, and a line AHB has been drawn cutting off a recipient ratio of OF to GB: so line AHB is positioned, because it resembles the first case of the seventh locus. And there is no limitation on it. And for that reason the synthesis of it is clear. And that is what we wanted to show.

Case 2. So let a line AB be drawn according to the second case, cutting off a recipient ratio of CA to BG. Let things be the same as before. So the ratio of CA to OF is a recipient. So the ratio of OF to BG is a recipient. So line AB is positioned, because it resembles the second case of the seventh locus. And that is what we wanted to show.

And we will find that as follows: Let there be taken a mean proportional between the two lines GF & FK, namely, line FB. And let BH be joined and produced in a straight line to point A. Then I say that line AB will cut off the ratio of CA to BG, which is less than all the ratios that the lines drawn from point H cut off, while cutting KG.

Let another line be drawn, such as NHD. Then, since line FB is a mean proportional between the two lines GF & FK, the ratio of OF to BG will be less than the ratio of SF to GN. So, by alternation, the ratio of OF to SF will be less than the ratio of BG to NG. However the ratio of OF to FS is like the ratio of

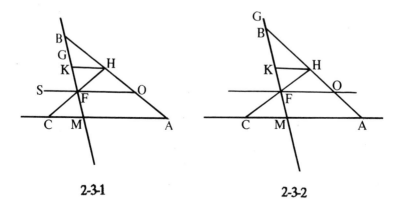

2-3-1   2-3-2

AC to CD. So the ratio of AC to CD is less than the ratio of
BG to NG. And by alternation, the ratio of AC to BG will be
less than the ratio of CD to NG. So line AB will cut off a ratio
of CA to BG, which is less than all the ratios that the lines
drawn through point H cut off, while cutting line KG. And that
is what we wanted to show.

And this problem will be synthesized thus: Let the rest be
as it was, and let there be taken a mean proportional between
the two lines GF & FK, namely, line FB. And let BH be joined
and produced in a straight line to point A. So line AB will cut
off a ratio of CA to BG less than all the ratios that the lines
drawn from point H cut off, while cutting KG.

And for that reason, when the ratio being given in the
synthesis is the same as the ratio of CA to BG, the line AB
alone will do what the problem requires.

And if the ratio is less than it, the problem will not be
synthesized.

But if it is greater than it, then it will be synthesized in two
ways, on each side of AB. So let the given ratio be the ratio of
P to T, which is greater than the ratio of CA to BG. And let it
be made that, as the ratio of CH to HF, so the ratio of P to J.
So, by way of equality, it will be clear that the ratio of J to T

will be greater than the ratio of FO to BG. However the ratio of SF to NG is greater than it, because line FB is a mean proportional between the two lines GF, FK. So it has become clear that two lines can be drawn from point H cutting off from the two lines GK, FO a ratio the same as the ratio of J to T, namely, on both sides of AB. And it is obvious from what has been supposed that the lines drawn in this way will do what the problem requires. And that is what we wanted to show.

Case 3. And let a line be drawn according to the third case, cutting off a recipient ratio of EC to AG. However the ratio of EC to DF is a recipient. So the ratio of DF to AG is a recipient. So line HE is positioned, because it resembles the third case of the seventh locus, which has no limitations. So the synthesis has become clear. And that is what we wanted to show.

Case 4. And let a line HE be drawn according to the fourth case, cutting off a recipient ratio of CE to AG. However the ratio of CE to FD is a recipient. So the ratio of FD to AG is a recipient. So line HE is positioned, because it resembles the fourth case of the seventh locus. And that is what we wanted to show.

And we will find that as follows: Let things be the same as they were, and let the mean proportional between the two lines GF, FK be either less than line FM or not less. So first let it be not less than it. And let HM be joined. Then I say that line HM cuts off a ratio of CM to MG greater than all the ratios that the lines drawn from point H cut off, while cutting line CM.

For let another line be drawn, such as HS. Then, since the mean proportional between the two lines GF, FK is not less than line FM, line HM will either cut off a greatest ratio, namely, the ratio of PF to MG, or it will be closer to the line which cuts off the greatest ratio. The ratio of PF to MG will be greater than the ratio of LF to SG. However the ratio of PF to LF is like the ratio of MC to CN. So the ratio of MC to CN is greater than the ratio of MG to SG. So, by alternation, the ratio of CM to MG will be greater than the ratio of CN to SG. So the

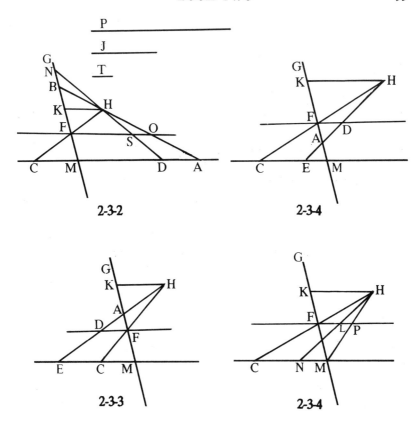

2-3-2

2-3-4

2-3-3

2-3-4

line HM will cut off the ratio of MC to MG, which is greater than all the ratios that the lines drawn from point H cut off, while cutting line CM. And that is what we wanted to show.

But let the mean proportional between the two lines GF, FK be less than line FM, namely, line FA. And let HA be joined and produced to E. Then I say that line HE will cut off the ratio of CE to AG greater than all the ratios cut off by lines drawn from point H, while cutting CM.

For, as for the lines which cut CM, line HM will cut off a ratio of CM to MG less than all of them. So let two lines be drawn, such as HX, HB. So, since line FA is a mean proportional between two lines GF & FK, line CA will cut off the ratio of FD to AG greater than all the ratios. So the ratio of

DF to AG is greater than the ratio of SF to GO. And by alternation, the ratio of DF to FS is greater than AG to GO. However the ratio of DF to FS is like the ratio of EC to CB, and is larger than the ratio of AG to GO. So the ratio of CE to AG is greater than the ratio of CB to OG. So likewise it also will be clear that it will cut off a ratio greater than CX to RG. So line HE will cut off a ratio of CE to AG greater than all the ratios that the lines drawn from point H cut off, while cutting CM.

And I say that line HM will cut off a ratio of CM to MG which is less than all the ratios that all the lines that cut line EM alone cut off. For, since line HX is nearer to the line HE, which cuts off the greatest ratio, than HM is, and the line closer to the line which cuts off the greatest ratio will cut off a greater ratio, so the ratio of line FP to line RG is greater than the ratio of LF to MG. And by alternation, the ratio of PF to FL is greater than the ratio of RG to GM. However the ratio of PF to FL is like the ratio of XC to CM. So the ratio of XC to CM is greater than the ratio of RG to MG. And by alternation, the ratio of CX to RG will be greater than the ratio of CM to MG. So the line HE will cut off the ratio of CE to AG greater than all the ratios that the lines drawn from point H cut off, while cutting CM. So, as for the line that cuts EM, HM will cut off a ratio less than the ratio that all these lines cut off. And that is what we wanted to show.

And this problem will be synthesized thus: Let the rest remain as it was. So the line that is a mean proportional between the two lines GF, FK will be either shorter than line MF, or not less than it. Then first let it not be less than it. And let HM be joined. So line HM will cut off the ratio of CM to MG, which is greater than all the ratios that lines drawn from point H cut off, while cutting CM.

And for that reason, if the ratio being given for the synthesis is the same as the ratio of CM to MG, then line HM alone will do what the problem requires.

And if the ratio is less, then the problem will be synthesized in one way. So let the given ratio be the ratio of P to T, which is less than the ratio of CM to MG. And let it be

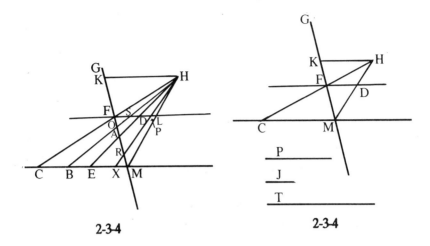

2-3-4   2-3-4

made that, as CH to HF, so P to J. So, by way of equality, it is
clear that the ratio of J to T will be less than the ratio of LF to
MG. And it will be clear that two lines can be drawn through
point H cutting off from two lines CM, MG a ratio the same as
the ratio of J to T. And when they have been drawn, they will
fall on either side of HM. So it is clear that the line which
passes through point H and cuts CM will do what the problem
requires. And for that reason, it will not obtain except in one
way. And that is what we wanted to show.

But let the line which is a mean proportional between the
two lines GF, FK be less than line FM, namely, line FN. And let
the two lines HM, HN be joined. And let line HN be produced
in a straight line to point S. So line HS will cut off a ratio of
CS to NG greater than all the ratios that lines drawn from
point H cut off, while cutting CM. And line HM will cut off
CM to MG which is less than all the ratios that lines drawn
through point H cut off, while cutting SM alone.

So for that reason, if the ratio which has been given for the
synthesis is the same as the ratio of CS to NG, it is clear that
line HS alone will do what the problem requires.

And if the given ratio is greater, then the <ratio> will not
be synthesized.

And if the ratio is less than the ratio of CS to NG but greater than the ratio of CM to MG, then it is also clear from what we have previously delimited that two lines will do what the problem requires, while cutting the two lines CS, SM on either side of HS.

And if the ratio is the same as the ratio of CM to MG, then it will also be clear from the premised limitation that the problem will obtain in two ways, with line HM, and with the other line that cuts CS.

But if the ratio is less than the ratio of CM to MG, then the line will fall outside of line CM while cutting off and will not do what the problem requires.

And everything will be clear according to the previously treated model. And that is what we wanted to show.

Case 5. And let a line HA be drawn according to the fifth case, cutting off a recipient ratio of CD to AG. However the ratio of CD to FB is a recipient. So the ratio of BF to AG is a recipient. And line HA is positioned, because it resembles that fourth case of the seventh locus which limits this. And that is what we wanted to show.

And we will find that as follows: The mean proportional between the two lines GF, FK is either greater than line FM or less. So first let it not be greater. When HM is joined, it will be clear, as we have previously delimited, that line HM will cut off a ratio of CM to MG which is greater than all the ratios that lines drawn from point H cut off, while cutting OM.

But let the mean proportional between the two lines GF, FK be greater than line FM, namely, line FL. When HL, HM are joined, it will be clear, as we have previously delimited, that line HL will cut off a ratio of CD to LG greater than all the ratios that lines drawn from point H cut off, while cutting OM. And line HM will cut off a ratio of CM to MG which is less than all the ratios that lines drawn from point H cut off, while cutting line LM alone. And that is what we wanted to show.

And the problem will be synthesized thus: With everything

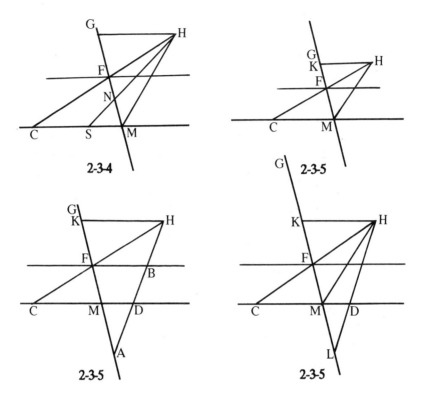

2-3-4

2-3-5

2-3-5

2-3-5

the same as before, let the line that is a mean between the two lines GF, FK be either greater than FM, or not greater than it.

So first let it not be greater than it. And let HM be joined. Then HM will cut off a ratio of CM to MG which is greater than all the ratios that lines drawn from point H cut off, while cutting OM.

And for that reason, if the ratio being given for the synthesis is the ratio of CM to MG, then HM alone will do what the problem requires.

And if it were greater than it, then what the problem requires will not be done.

And if the ratio is less than it, it will be clear from the things introduced before that only one line will be drawn which will cut OM while doing what the problem requires. And that is what we wanted to show.

But let the mean proportional between the two lines GF, FK be longer than line FM, namely, line FL. And let HM, HL be joined. Then line HL will cut off a ratio of CD to LG which is greater than all the ratios that lines drawn from point H cut off, while cutting OM. And line HM will cut off a lesser ratio, namely, the ratio of CM to MG.

So, if the ratio in the synthesis has been given the same as the ratio of CD to LG, it will be clear that the line HL alone will do what the problem requires.

And if the ratio is greater, then there will be no synthesis.

And if it is less than the ratio of CD to LG but greater than CM to MG, it is clear, from what has been said previously, that the problem will obtain in two ways, namely, while cutting LM on each side of HL.

And if it is less than the ratio of CM to MG, then it will also be clear, from the preceding limitations, that one line only will do what the problem requires, one that cuts OL.

And if the ratio is the same as the ratio of CM to MG, then the problem will obtain in two ways, because line HM will do what the problem requires, and also another line which cuts LO.

And all that we have stated will be clear most easily from what has preceded. And that is what we wanted to show.

Let the line drawn from H parallel to line AB fall above point G as line HK. That is, point G will be between it and point M. And when a straight line has been drawn from H to point G, it will either pass through point C; or between it and point A; or between it and point M.

Locus 4. So first let it fall above C, that is, between it and point A, as line HS. Furthermore it is clear that lines drawn from point H will fall in five positions.

Case 1. So let a line HN be drawn according to the first case, cutting off a recipient ratio of CN to GP. And let line SL be drawn through point S parallel to line MP. And let NP be produced in a straight line to point L. So, since point G is a

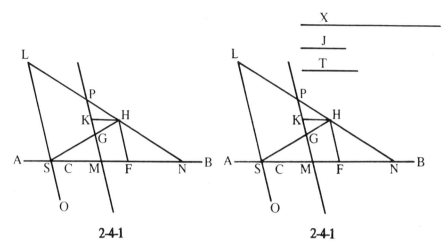

2-4-1   2-4-1

recipient, line HS is positioned. And line AB is positioned. So point S is a recipient. So line OSL is a recipient. And the ratio of SH to HG is a recipient. And since the ratio of CN to GP is a recipient, while the ratio of GP to LS is a recipient, because it is like the ratio of GH to HS: so the ratio of CN to SL is a recipient. And since there are two recipient lines, namely, SL & AB, and the terminal point on line SL is at S while the one on line AB is at C, and the recipient point H is within angle LSB, and the parallel line falls above point C, and NHL has been drawn, cutting off a recipient ratio of SL to CN: so line HL is positioned, since it resembles the first case of the sixth locus, in which there is no restriction. And that is what we wanted to show.

And this problem will be synthesized as follows: Things will be the same as before. And let the given ratio be the ratio of X to T. And let it be made that, as the ratio of HG to HS, so the ratio of X to J. And since there were two recipient lines, namely, lines OL & AB, with line OL passing through point S and line AB passing through point C, and the recipient point H was within angle LFB, and the line that was drawn through point H parallel to line LO cuts off above point C, and the ratio was the ratio of J to T: so let LH be drawn according to the first case of the sixth locus, cutting off a recipient ratio of SL to CN the same as the ratio of J to T.

Then, since the ratio of GH to HF, that is, GP to SL, is like
the ratio of X to J, while the ratio of SL to CN is like the ratio
of J to T, so, by way of equality, the ratio of GP to CN will be
like the ratio of X to T. So line HL will do what the problem
requires, and that is what we wanted to show.

Case 2. Now let a line HL be drawn according to the second
case, cutting off a recipient ratio of CN to GL. And let HG be
produced to point S, and let a line OSP be drawn parallel to
line ED. So, since each of the two points H, G is a recipient,
line HS will be a recipient. And line AB is positioned. So point
S is a recipient. But line OSP has been drawn through a
recipient point S parallel to line ED. So line OP is a recipient.
And the line HF, which is parallel to line ED and passes
through point H and meets AB at F, is a recipient. And let HL
be produced to P. So, since the two points S, H are recipient,
the ratio of SH to HG is a recipient. And for that reason, the
ratio of PS to LG is recipient. But the ratio of LG to CN is a
recipient. So, since the two lines OP, AB are recipient, and OP
has a terminal point at S, and AB passes through point C, and
the parallel line HF cuts beyond the recipient point C, and line
HP was drawn according to the second case of the sixth locus
treated in the first book, cutting off a recipient ratio of SP to
CN: so line HP is positioned. And it has a limitation.

Since the line that is a mean proportional between the two
line FS, SC is either greater than SM, or not greater than it,
first let it be not greater than it. And let HM be joined. Then I
say that line HM will cut off a ratio of CM to MG greater than
all the ratios that the lines drawn from point H cut off, while
cutting line FM <in the direction of point D.> Now, since the
line which is a mean proportional of the two lines FS, SC is not
greater than line SM, it is either equal to it or less than it.

So, if it is equal to line SM, while the two lines OP, AB are
recipient, and the terminal point on OP is at point S, and the
terminal point on line AB is at point C, and the parallel line
falls beyond the position of point C as line HF, <when HM is
produced in a straight line to R> according to the second case
of the sixth locus: then it will cut off a ratio of RS to CM less

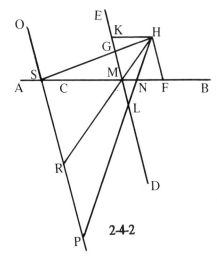

2-4-2

than all the ratios that the lines thus drawn cut off. [And if the mean proportional is less than line SM, HM] will be nearer to the line that cuts off the least ratio than line HP is. So for that reason the ratio of line RS to line CM is less than the ratio of PS to CN. And by alternation, the ratio of RS to SP, which is like the ratio of MG to GL, will be less than the ratio of MC to CN. And by alternation, the ratio of GM to MC will be less than the ratio of GL to CN. And by inversion, the ratio will be the ratio of CM to MG, greater than the ratio of CN to GL. And likewise also it will be clear that line HM will cut off ratios greater than all the ratios that the drawn lines cut off, while cutting line MF. So line HM will cut off a ratio of CM to MG which is a greatest ratio. And that is what we wanted to show.

And let the line which is the mean proportional between the two lines FS, SC be greater than line SM, and let the equal to it be line NS. Let HN be joined and produced in a straight line to L. And let HM be joined. Then I say that line HL will cut off a ratio of CN to GL which is greater than all the ratios that lines drawn through point H cut off, while cutting all along MD.

For let HM, HL be produced to the two points R, P. And let two lines be drawn, on each of the two sides of HP, such as HZ, HX. Then, since the two lines OP, AB are positioned, and

OP <passes through> point S while AB <passes through> point C, and the line parallel to ORP, which passes through the recipient point H beyond point C, is line HF, and line SN is the mean proportional between lines FS & SC, and HN has been produced in a straight line to point P, according to the second case of locus six, cutting off the least ratio of PS to CN, and another line, such as line HX, has been drawn: so the ratio of line PS to CN is less than the ratio of XS to CT. And by alternation, the ratio of PS to XS is less than the ratio of CN to CT. However the ratio of PS to XS is like the ratio of LG to GJ. So the ratio of LG to GJ is less than the ratio of NC to CT. And by alternation, the ratio of LG to CN is less than the ratio of GJ to CT. So the ratio of CN to LG is greater than the ratio of CT to GJ. So line HL will cut off a ratio which is greater than the ratio that line HJ cuts off, namely, the ratio of CN to LG, and greater than all the ratios which all the lines thus produced cut off, while cutting all along line MD.

And we say that line HM will cut off a ratio of CM to MG which is less than all the ratios that lines drawn from point H cut off, while cutting line ML alone. For, since according to the second case of the sixth locus, HP cuts off a ratio of PS to CN, which is least, and likewise it will always be the case that lines near line HP cut off ratios less than the ratios that lines distant from it cut off, so the ratio of line ZS to line CU is less than the ratio of RS to CM. And by alternation, the ratio of ZS to RS will be less than the ratio of UC to CM, while the ratio of ZS to SR is like the ratio of GV to GM. So the ratio of VG to MG is less than the ratio of UC to CM. And by alternation, the ratio of GV to UC is less than the ratio of MG to CM. And by inversion, the ratio of CU to VG is greater than the ratio of CM to MG. So line HM will cut off a ratio less than the ratio that line HV cuts off. And for that reason it is clear that line HM will cut off the ratio of CM to MG which is less than all the ratios that lines drawn from point H cut off, while cutting line ML. And that is what we wanted to show.

And this problem will be synthesized thus: Let things be the same as before, and first let the mean proportional between lines FS, SC be not greater than line SM. And let HM be joined.

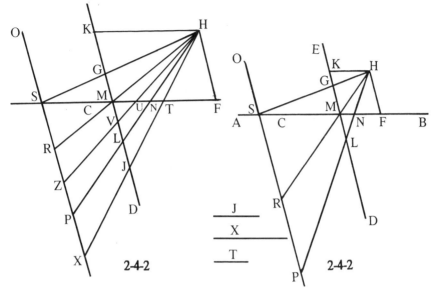

2-4-2   2-4-2

So line HM will cut off a ratio of CM to MG which is greater
than all the ratios that all the lines cut off. And the ratio given
in the synthesis will either be the same as the ratio of CM to
MG, or greater than it, or less.

So, if it is the same as it, then line HM will do what the
problem requires. And if the ratio is greater, then the problem
will not be synthesized, since it was given greater than the
greatest ratio. And if the ratio is less, then the problem will
obtain in one way.

For let the given ratio be the ratio of J to T, which is less
than the ratio of CM to MG. And let it be made that, as the
ratio of SH to HG, so the ratio of X to T. And let HM be
produced in a straight line. So it is clear from the limitation
that the ratio of RS to CM, according to case two of locus six
of book one, is either the least, or is the nearest to the least
ratio for all the lines that are drawn, while cutting RP. And
since the ratio of SH to HG is like the ratio of X to T, and like
the ratio of RS to MG, so the ratio of X to T is like the ratio of
RS to MG. But the ratio of T to J is greater than the ratio of
GM to MC. So, by way of equality, the ratio of X to J will be
greater than the ratio of RS to CM. And since the ratio of RS

to CM is either the least in the second case of the sixth locus, or less than all the ratios that the drawn lines cut off, while cutting RP, and the ratio of X to J is greater than it, so two lines will be drawn, on both sides of line HM, cutting off a ratio the same as the ratio of X to J. One of these will not do what the problem requires, namely, the line that is drawn cutting line MG. But the other line will do what the problem requires, namely, the line that is drawn cutting MD. For let line HP be drawn cutting off the ratio of PS to CN the same as the ratio of X to J. Then I say that line HP will do what the problem requires. That is, that the ratio of CN to GL is like the ratio of J to T.

For, since the ratio of SH to HG is like the ratio of X to T, and like the ratio of PS to LG, the ratio of X to T will be like the ratio of PS to LG. So, since the ratio of J to X is like the ratio of CN to SP, and the ratio of X to T like the ratio of PS to LG, so, by way of equality, the ratio of J to T is like the ratio of CN to LG. So line HLP will do what the problem requires. And it is clear that it alone does so, and this is what we wanted to show.

And let the line which is a mean proportional between lines FS, SC be greater than line SM, namely, line SN. And let the two lines HN, HM be joined and produced in a straight line to the two points P, R. So line HP will cut off a ratio of CN to LG greater than all the ratios that lines drawn through point H cut off, while cutting the two lines CB, GD. And line MH will cut off the ratio of MC to MG less than all the ratios that lines drawn from H cut off, while cutting line MN. So the given ratio will be either the same as the ratio of NC to GL, or greater than it, or less than it but greater than the ratio of CM to MG, or the same as this, or less.

So if it is the same as the ratio of NC to LG, line LH alone will do what the problem requires.

And if it is greater, then the problem will not be synthesized.

And if it is less but greater than the ratio of CM to MG, then the problem will be synthesized twice, on both sides of line PH.

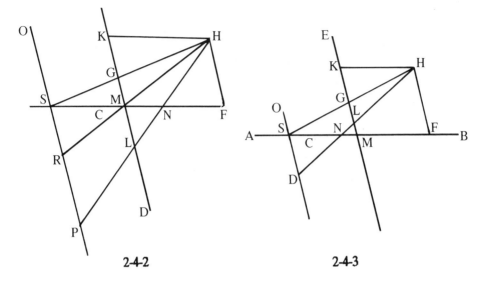

2-4-2   2-4-3

And if it is the same as the ratio of CM to MG, the problem will be synthesized twice also. For CM will do what it requires and <FN will be cut off> on the other side of LH.

And if it is less, then it will be synthesized once, cutting from the side of LH which is toward D.

Case 3. And let a line HL be drawn according to the third case, cutting off a recipient ratio of NC to LG. However the ratio of LG to DS is a recipient. So the ratio of DS to NC is a recipient. So line HL is positioned, as it resembles a case which was in the first book, namely, the second case of the sixth locus.

And we will find that as follows: Since the line which is the mean proportional between lines FS, SC is either less than line SM or not less than it, then first let it be not less than it.

And let line HM be joined and produced in a straight line to point R. And it will be clear, as we have previously shown, that line MH will cut off a ratio of CM to MG which is greater than all the ratios that lines drawn through point H cut off, while cutting all along line CM. And that is what we wanted to show.

But let the line which is the mean proportional of the two lines FS, SC be less than line SM, namely, line SN. Let HN be drawn and let HM be joined and let them both be produced to the two points D, R. And it will be clear from the limitations previously explained that line HN will cut off a ratio of CN to LG which is greatest of all the ratios that lines drawn from point H cut off, while cutting CM. As for the lines that are drawn cutting line NM alone, line HM cuts off a ratio less than the ratios that they cut off, namely, the ratio of CM to MG. And that is what we wanted to show.

And the problem will be synthesized thus: With the rest as before, let the line which is the mean proportional between FS, SC either be less than line SM, or not less than it.

So first let it be not less than it. And let HM be joined and produced to point R. So line HMR will cut off a ratio of CM to MG which is greater than all the ratios which lines drawn through point H cut off, while cutting off line MC.

And for that reason, when the ratio in the synthesis has been given the same as the ratio of CM to MG, it will be clear that line HM alone will do what the problem requires.

But if the ratio is given greater than it, the problem will not be synthesized.

And if it has been given less, then it will be clear from the preceding synthesis that one line will do what the problem requires, while cutting line CM.

But let the mean proportional line between the two lines SF, SC be less than line SM, namely, line SN. And let the two lines HM, HN be joined and produced to the points R, D. So line HD will cut off a ratio of CN to LG which is greater than all the ratios that lines drawn from point H cut off, while cutting all along line CM. But as for the lines that cut NM alone, line HM will cut off a ratio CM to MG which is less than all the ratios that those lines cut off.

And for that reason, when the ratio given in the synthesis is the same as the ratio of CN to LG, it will be clear that line HN will do what the problem requires.

And if the ratio has been given greater than it, the problem

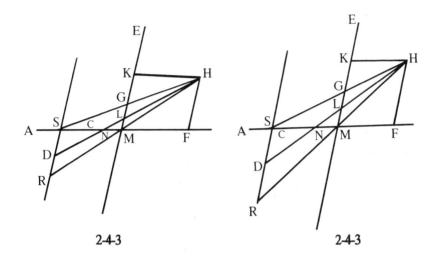

2-4-3   2-4-3

will not be synthesized.

And if it has been given less than the ratio of CN to LG but greater than the ratio of CM to MG, then it will be clear from the previously stated restriction that the problem will be synthesized in two ways, on both sides of HN, cutting CN, NM.

And if the given ratio is less than the ratio of CM to MG, then it will also be clear from the preceding limitations that one line will do what the problem requires while cutting line CN.

And if the given ratio is the same as the ratio of CM to MG, it will also be clear from the preceding limitations that the problem will obtain in two ways. And to that end, line HM will do what the problem requires, and the other line will cut off NC, and that is what we wanted to show.

Case 4. And let a line HLN be drawn according to the fourth case, cutting off a recipient ratio of CN to LG. And let it be produced to point D. However the ratio of LG to DS is a recipient. So the ratio of DS to NC is a recipient. So line HLN is positioned, because it resembles the third case of the sixth locus, which has no limitation. And it will be synthesized accordingly. So what we wanted to show has been proved.

Case 5. And let a line HLN be drawn according to the fifth case, cutting off a recipient ratio of LG to NC. However the ratio of LG to SB is a recipient. So the ratio of SB to NC is a recipient, and line HL is positioned, because it resembles the fourth case of the sixth locus. And that is what we wanted to show.

We will find that as follows: With the rest as before, let SN be a mean proportional between lines FS, SC. And let HN be joined. Then I say that line HLN will cut off a ratio of GL to NC which is greater than all the ratios that lines drawn through point H cut off, while cutting all along line SA.

For let another line be drawn, such as line HX. So, since line SN is a mean proportional between FS & SC, the ratio of SB to NC is greater than the ratio of SR to XC. And by alternation, the ratio of BS to SR is greater than the ratio of NC to CX, while the ratio of BS to SR is like the ratio of GL to GP, so the ratio of GL to GP is greater than the ratio of NC to CX. And by alternation, the ratio of LG to NC is greater than the ratio of PG to XC. So line HLN will cut off a ratio of GL to CN which is the greatest cut off by a line from point H, while cutting line SA. And that is what we wanted to show.

And this problem will be synthesized as follows: With the rest as before, let line SN be made the mean proportional between FS & SC, and let NH be joined. Then line NH will cut off a ratio of GL to NC which is the greatest of all the ratios that the lines drawn from point H cut off, while cutting all along SA.

And for that reason, if the ratio given in the synthesis is the same as the ratio of LG to NC, then line HLN alone will do what the problem requires.

But if the given ratio is greater than it, then the problem will not be synthesized.

And if it is less than it, it will be clear from has been stated that two lines will do what the problem requires while cutting the two lines AN, NS. And that is what we wanted to show.

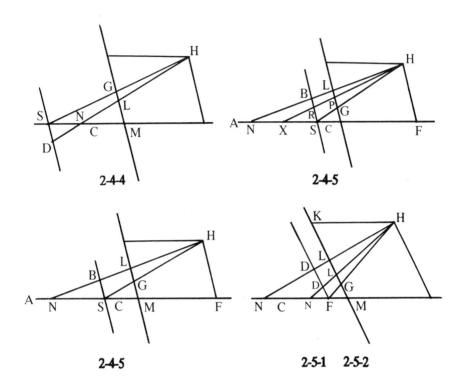

2-4-4

2-4-5

2-4-5

2-5-1   2-5-2

Locus 5. And also let the line HF, drawn from point H through point G, fall below point C. Then it will be clear that the lines drawn from point H will fall in five positions.

Case 1 & 2. So let a line NH be drawn according to the first and second cases, cutting off a recipient ratio of GL to NC. However the ratio of GL to FD is a recipient. So the ratio of FD to NC is a recipient. So, since there are two positioned lines DF & CM, cutting each other at F, which is the terminal point on line DF, and the terminal point on line NM is point C, and the recipient point H is within angle DFM, and a line HN has been drawn according to the first and second cases, cutting off a ratio of FD to NC: then line HN is positioned, since it resembles the first and second cases of the fourth locus, in which there is no limitation.

And this problem will be synthesized thus: With things being as before, let the given ratio be the ratio of N to O. And let it be made that, as the ratio of HG to FH, so the ratio of N to J. And since there are two positioned lines FD & CM, and the point which is passed by on line FD is at point F, and the point which is passed by on line CM is at point C, and the recipient point H is within angle MFD: then, let a line NH be drawn according to the first and second cases of the fourth locus, cutting off a ratio of FD to NC the same as the ratio of J to O. So it is clear that line HN will do what the problem requires, and that is what we wanted to show.

Case 3. Now let a line HN be drawn according to the third case, cutting off a recipient ratio of GL to NC. And let it be produced to point E. However the ratio of GL to EF is a recipient. So the ratio of EF to CN is a recipient. So line HNE is positioned, as has been explained in the third case of the fourth locus. And that is what we wanted to show.

And we will find it thus: Let HM be joined and produced in a straight line to point B. Then I say that line HB will cut off a ratio of GM to MC which is greater than all the ratios that the lines drawn from point H cut off. For let another line be drawn, such as line HE. Then, since it is clear that the line nearer to point F will always cut off a ratio less than the ratio that the line more remote from it cuts off, the ratio of EF to CN will be less than the ratio of BF to CM. So, by alternation, the ratio of BF to FE will be greater than the ratio of MC to NC. However the ratio of BF to FE is like the ratio of MG to GL. So the ratio of MG to GL is greater than the ratio of MC to CN. And by alternation, the ratio of MG to MC is greater than the ratio of LG to CN. So line HB will cut off a ratio of MG to MC which is greater than all the ratios that the lines drawn through point H cut off, while cutting line FM.

And this will be synthesized thus: Let the rest be as before, and let HM be joined and produced in a straight line to point B. So line HB will cut off a ratio of GM to MC which is greater than all the ratios that the lines drawn through point H cut off,

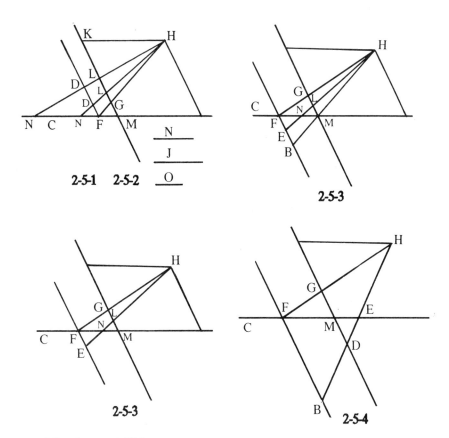

2-5-1   2-5-2

2-5-3

2-5-3

2-5-4

while they cut FM.

So, if the ratio being given in the synthesis is the same as the ratio of MG to MC, the line HB alone will do what the problem requires. And if the ratio is greater than it, then the problem will not be synthesized. And if the ratio is less than it, it will be clear from the prevously stated limitations that the problem will obtain in one way, while the line drawn cuts line FM. And that is what we wanted to show.

Case 4. And let a line HD be drawn according to the fourth case, cutting off a recipient ratio of GD to CE, and let HD be produced to point B. However the ratio of GD to BF is a recipient. So the ratio of BF to CE is a recipient. So line HB is positioned, because it resembles the same case as the preceding.

And we will find that in this way: Let the rest be as before, and let line HM be joined and produced to point A. Then I say that line HA will cut off a ratio of GM to MC which is less than all the ratios that the lines drawn from point H cut off, while cutting line JM. For let another line be drawn, such as HB. Then, since the lines that are nearer to point F cut off ratios less than the ratios that the lines more distant from it cut off, so the ratio of FA to CM is less than the ratio of BF to CE. And by alternation, the ratio of FA to FB is less than the ratio of MC to CE. But the ratio of AF to FB is like the ratio of MG to GD. So the ratio of MG to GD is less than the ratio of CM to CE. So, by alternation, the ratio of GM to MC is less than the ratio of GD to EC. So line HA will cut off a ratio of GM to MC which is less than all the ratios that the lines drawn through H cut off, while cutting line JM, and that is what we wanted to show.

And this problem will be synthesized thus: Let line HM be joined and produced in a straight line to point A. So the line HA will cut off a ratio of GM to MC which is less than all the ratios that the lines drawn from point H cut off, while cutting line JM.

And if the ratio being given in the synthesis is the same as the ratio of GM to MC, then line CA alone will do what the problem requires. And if it is less than it, the problem will not be synthesized. But if it is greater, it will be clear from what has been explained that one line will do what the problem requires while cutting line MJ. And that is what we wanted to show.

Case 5. Now let a line HA be drawn according to the fifth case, cutting off a recipient ratio of GB to CA. And let it be produced to point D. However the ratio of GB to FD is a recipient. So the ratio of FD to CA is a recipient. So line AD is positioned, because it resembles the fourth case of the fourth locus, in which there is no restriction. And the synthesis of this problem is clear.

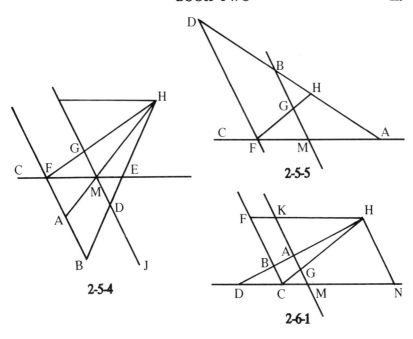

2-5-5

2-5-4

2-6-1

Locus 6. And let the line drawn from point H to point G fall on point C, namely, line GC. Then it is clear that, <reckoned from that line,> the lines drawn from point H will fall in four positions.

Case 1. So first let a line <HD> be drawn according to the first case, cutting off a recipient ratio of GA to CD. <Let it cut line CB at B.> However the ratio of GA to BC is a recipient. So the ratio of BC to CD is a recipient. So line BD is positioned, as has been made clear in the first case of the third locus.

And it will be necessary that the ratio being given in the synthesis be less than the ratio of KG to CN. For let the line drawn parallel be line HN, and let the two lines CB, HK be drawn to point F. Then, since the ratio of FC to CB is greater than the ratio of FB to BC, while the ratio of FC to CB is like the ratio of KG to GA, and the ratio of FD to BC is like the ratio of FH to CD, while line FH is equal to line CN, so the ratio of KG to GA is greater than the ratio of CN to CD. And by alternation, the ratio of KG to CN will be greater than the ratio of GA to CD. So it will be necessary that the given ratio be less than the ratio of KG to CN.

And the problem will be synthesized thus: Let the given ratio be the ratio of K to L, which is less than the ratio of KG to CN. And let it be made that, as the ratio of GH to HC, so the ratio of K to O. However the ratio of GH to CH is like the ratio of GK to KM. So the ratio of GK to KM is like the ratio of K to O. However line KM is equal to line HN. So the ratio of GK to HN is like the ratio of K to O. And by inversion, the ratio of O to K will be like the ratio of NH to KG. However the ratio of K to L is less than the ratio of KG to CN. So the ratio of O to L is less than the ratio of HN to NC. So, if it is made that, as the ratio of O to L, so the ratio of HN to another line, then that line will be greater than line CN. So let it have that ratio to line ND and let HD be joined. Then I say that line HD will do what the problem requires.

For, since the ratio of K to O is like the ratio of GH to CH, while the ratio of GH to CH is like the ratio of GA to BC, so the ratio of K to O is like the ratio of GA to CB. And the ratio of O to L is like the ratio of BC to CD, that is, NH to ND. So the ratio of K to L will be like the ratio of GA to CD. So line HD will do what the problem requires. And that is what we wanted to show.

Case 2. And let a line HB be drawn, cutting off a recipient ratio of BG to CA. And let it be produced to point D. However the ratio of BG to CD is a recipient. So the ratio of CD to CA is a recipient. So line HD is positioned, because it resembles the second case of the third locus.

And we find that in this way: Let HM be joined and produced in a straight line to point E. Then I say that line HM will cut off a ratio of GM to MC which is less than all the ratios that the lines drawn through point H cut off, while cutting all along line FM. For let HD be drawn. And since the lines near to point C always cut off ratios less than the ratios that the lines distant from it cut off, as has been made clear, so the ratio of line EC to CM is less than the ratio of DC to CA.

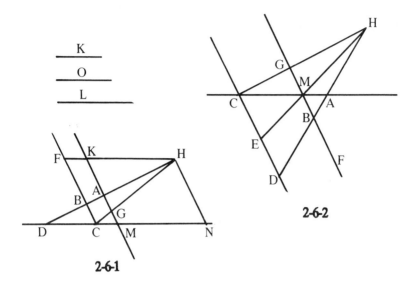

2-6-2

2-6-1

And by alternation, the ratio of EC to CD is less than the ratio of MC to CA. However the ratio of EC to CD is like the ratio of MG to GB. So the ratio of MG to GB is less than the ratio of MC to CA. And by alternation, the ratio of MG to MC is less than the ratio of GB to CA. So line CM will cut off a ratio of GM to MC which is less than all the ratios that the lines drawn from point H cut off, while cutting all along line FM.

And this problem will be synthesized as follows: Let the rest be as before, and let HM be joined and produced to point E. So line HM will cut off a ratio of GM to MC which is less than all the ratios that the lines drawn from point H cut off, while cutting all along line FM.

And for that reason, if the ratio being given in the synthesis is the same as the ratio of GM to MC, then line HM alone will do what the problem requires. And if the given ratio is less, then the problem will not be synthesized. And if it is greater, it will be clear from what has been previously stated that the line will do what the problem requires while cutting off line FM. And this is what we wanted to prove.

Case 3. And let a line HB be drawn according to the third case, cutting off a recipient ratio of AG to BC. And let this be produced in a straight line to point D. However the ratio of AG to CD is a recipient. So the ratio of BC to CD is a recipient. So line HA is positioned, because it resembles the case cited before.

And we will find it thus: Let HM be joined and produced in a straight line to point E. So, since lines near to point M always cut off ratios greater than the ratios that lines distant from it cut off, it will be clear from the previously stated limitations that line HM will cut off a ratio of GM to MC, which is greater than all the ratios that the lines drawn from point H cut off, while cutting all along line MC.

And the problem will be synthesized thus: With the rest the same as before, let HM be joined and produced to point E. So line HM will cut off a ratio of MG to MC, which is greater than all the ratios that the lines drawn through point H cut off, while cutting all along line MC.

And for that reason, if the ratio being given in the synthesis is the same as the ratio of GM to MC, then line HM alone will do what the problem requires. And if the ratio is greater then the problem will not be synthesized. And if it is less, then, in accordance with all the previously stated limitations, it is clear that the line that does what the problem requires will cut line MC, and that is what we wanted to show.

Case 4. And let a line HB be drawn according to the fourth case, cutting off a recipient ratio of BG to AC, and let it be produced to point D. However the ratio of GB to CD is a recipient. And since two lines MA, CD have been positioned, and the point which is passed by on each of them is at point C, and point H is within angle ACD, and AD has been drawn according to the third case of the third locus, cutting off a recipient ratio of DC to CA: so line AD is positioned, as has been made clear in the third case of the third locus, in which there is no limitation. And that is what we wanted to show.

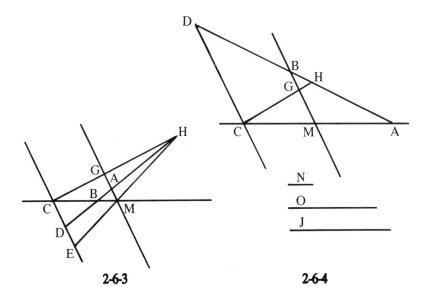

2-6-3   2-6-4

And this problem will be synthesized as follows: With the rest as before, let the given ratio be the ratio of N to J. And let it be made that, as the ratio of HG to CH, so the ratio of N to O. So, since there are two positioned lines, namely, lines MA & DC, and the point which is passed by on them both is at point C, and the recipient point H is within angle ACD: so let AD be drawn according to the third case of the third locus, cutting off a ratio of DC to CA the same as the ratio of O to J. So it will be clear that line AD will do what the problem requires.

And now let the point be within angle AMB as point H. And through point H let the two lines CH, HG be drawn parallel to AM, MB.

Locus 7. And first let them cut at the two terminal points on the lines, namely, at the two points C, G. Then it is clear that the lines drawn from point H will fall in three positions.

Case 1. So first let a line AB be drawn according to the first case, cutting off a recipient ratio of GB to AC. However the ratio of GB to CA is like the ratio of rectangle GB by CA to the square on CA. So the ratio of rectangle BG by CA to the square on CA is a recipient. But rectangle BG by CA is a recipient, since it is equal to the rectangle GH by CH. So line CA is a recipient. So point A also is a recipient. But point H is a recipient. So line AB is positioned. And that is what we wanted to show.

And the problem will be synthesized thus: With the rest as before, let the given ratio be the ratio of K to L, and let it be made that, as the ratio of K to L, so the ratio of rectangle GH by CH to the square on CA. Let AB be joined and produced in a straight line to point B. Then I say that line AB will do what the problem requires.

For, since the ratio of K to L is like the ratio of rectangle CH by HG to the square on CA, while rectangle GB by CA is equal to CH by HG, so the ratio of K to L is like the ratio of rectangle GB by CA to the square on CA. So the ratio of K to L is like the ratio BG to CA. So line AB will do what the problem requires. And that is what we wanted to show.

Case 2. And let a line HK be drawn according the the second case, cutting off a recipient ratio of CL to KG. Then, since the ratio of CL to KG is a recipient, the ratio of rectangle CL by KG to the square on KG is a recipient. However rectangle CL by KG is equal to rectangle CM by MG. So the ratio of rectangle CM by MG to the square on KG is a recipient. However the rectangle CM by MG is a recipient, since each of the two lines CM, MG is recipient. So the square on KG is a recipient. So line KG is recipient in magnitude and position. And point G is a recipient. So point K. So line KH.

And since line CM is greater than line CL while line MG is less than line GK, so the ratio of CM to CL is greater than the ratio of MG to GK. And by alternation, the ratio of CM to MG will be greater than the ratio of CL to KG. But the ratio of CL

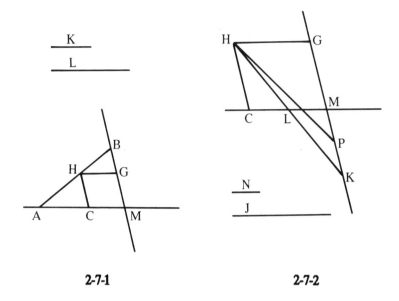

2-7-1   2-7-2

to KG is a recipient. So it will be necessary that the ratio being given in the synthesis be less than the ratio of CM to MG.

And this problem will be synthesized in this way: With the rest as before, let the given ratio be the ratio of N to J, which is less than the ratio of CM to MG. So, since the ratio of CM to MG is greater than the ratio of N to J, the ratio of rectangle CM by MG to the square on MG is greater than the ratio of N to J. So, if it be made that, as the ratio of N to J, so the ratio of rectangle CM by MG to another rectangle, then it will have that ratio to a rectangle greater than the square on MG. So let it be to the square on GK.

Then I say that line HK will do what the problem requires, that is, that the ratio of N to J will be like the ratio of CL to KG. For, since the ratio of N to J is like the ratio of rectangle CM by MG to the square on KG, while rectangle CM by MG is equal to rectangle CL by KG, the ratio of N to J will be like the ratio of rectangle of CL by KG to the square on KG, that is, the ratio of CL to KG. So the line KH will do what the problem requires. And I say that it alone does so. For let another line be drawn, such as line HP. Then it is clear that one of them will do what the problem requires, and the other will fail. And that is what we wanted to show.

Case 3. Let the rest be as before, and let line HK be drawn according to the third case, cutting off a recipient ratio of CL to KG. So, since the ratio of CL to KG is a recipient, the ratio of rectangle CL by KG to the square on KG is a recipient. However rectangle CL by KG is equal to rectangle CM by MG. So the ratio of rectangle CM by MG to the square on KG is a recipient. But rectangle CM by MG is a recipient, since each of the two lines CM, MG is a recipient. So the square on KG is a recipient. So line KG is recipient in magnitude and position. But point G is a recipient. So point K is a recipient. But point H is a recipient. So line HK is positioned.

And since line LC is longer than line LM, while line KG is less than GM, the ratio of line LC to LM is greater than the ratio of KG to GM. And by alternation, the ratio of CL to KG is greater than the ratio of LM to MG. However the ratio of CL to KG is a recipient. So it will be necessary that the ratio being given in the synthesis be greater than the ratio of LM to MG.

And the problem will be synthesized in this way: With the rest as before, let the given ratio be the ratio of N to J, which is greater than the ratio of LM to MG, and greater than the ratio of rectangle LM by MG to the square on MG. Then, if it is made that, as the ratio of N to J, so the ratio of rectangle LM by MG to some other rectangle, then it will have that ratio to a rectangle less than the square on MG. So let that be to the square on KG. And let HK be joined and produced in a straight line to point L. Then I say that line HL will do what the problem requires, that is, that the ratio of N to J is like the ratio of CL to KG.

For, since the ratio of N to J is like the ratio of LM by MG to the square on KG, while rectangle LM by MG is equal to

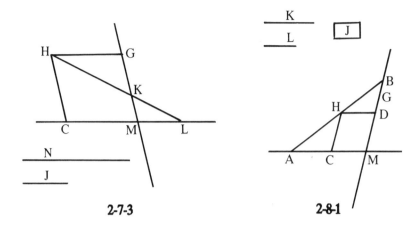

2-7-3   2-8-1

rectangle CL by KG, then the ratio of N to J is like the ratio of rectangle CL by KG to the square on KG, or the ratio of CL to KG. So line HL will do what the problem requires, and only it will be able to. For, if another line were drawn, one of these two will do what the problem requires, and the other will fail. And that is what we wanted to show.

Locus 8. Now let the line that is drawn through point H parallel to line GM cut at point C, and let the line that is drawn parallel to line MA fall short of point G as line HD. Then the lines that are drawn through point H fall in four ways.

Case 1. So first let a line AHB be drawn according to the first case, cutting off a ratio of CA to GB the same as a given ratio. And the ratio of AC to GB is like the ratio of rectangle AC by DB to rectangle DB by BG. So the ratio of rectangle AC by AB to rectangle DB by BG is a recipient. However the rectangle AC by DB is recipient, since it is equal to rectangle CH by HD. So rectangle DB by BG is a recipient. And it has been applied to the recipient line DG with the addition of the square on GB. So point B is a recipient, while point H is a recipient. So line AB is positioned.

And the problem will be synthesized thus: With the rest as

before, let the given ratio be the ratio of K to L, and let it be made that, as the ratio of K to L, so the ratio of rectangle CH by HD to rectangle J. And let there be applied to line DG a rectangle equal to rectangle J with the excess of a square, namely, rectangle DB by BG. And let BH be joined and produced in a straight line to point A. Then I say that line AB will do what the problem requires.

For, since the ratio of K to L is like the ratio of rectangle CH by HD to rectangle J, while rectangle J is equal to rectangle DB by BG, and rectangle CH by HD is equal to rectangle AC by DB, then the ratio of K to L is like the ratio of rectangle AC by DB to rectangle DB by BG, that is, like the ratio of AC to GB. So line AB will do what the problem requires. And that is what we wanted to show.

Case 2. And let a line AB be drawn according to the second case, cutting off a recipient ratio of AC to BG. So the ratio of rectangle AC by DB to rectangle DB by BG is a recipient, because it is the same as the ratio of AC to BG. However rectangle AC by DB is a recipient, because it is equal to rectangle CH by HD. So rectangle DB by BG is a recipient, and it has been applied to a line DG with the deficiency of a square. So line BD is a recipient. But D is a recipient. So B is a recipient. But point H also is a recipient. So line AB is positioned.

And since it is necessary that the ratio in the synthesis be the ratio of rectangle CH by HD to another rectangle, and that there be applied to line DG a rectangle equal to that rectangle with the deficiency of a square, while it is not possible for us to apply to every given line a rectangle equal to a <given> rectangle with the defect of a square: then for that reason it will not be possible to draw a straight line to a point A which cuts off <outside the two lines CH, HD a ratio> the same as any given ratio, which is what we wanted to show.

And we will find that as follows: Let the rest be as before. And let line DG be cut in half at point F, and let FH be joined and produced in a straight line to point E. Then I say that line

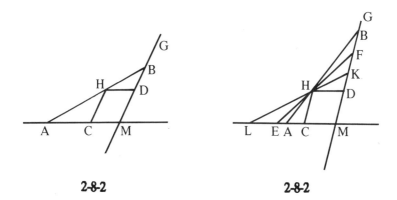

2-8-2   2-8-2

FE will cut off a ratio of EC to FG which is less than all the ratios that the lines drawn through point H cut off, while cutting all along line DG.

For let another line be drawn, such as line AB. So, since line DF is equal to line FG, the rectangle DF by FG will be greater than rectangle DB by BG. And since rectangle EC by DF is equal to rectangle AC by DB, because each of them is equal to rectangle CH by HD, while the ratio of rectangle EC by DF to rectangle DF by FG is less than the ratio of rectangle AC by DB to rectangle DB by BG, and while the ratio of rectangle EC by DF to rectangle DF by FG is like the ratio of EC to GF, and the ratio of rectangle AC by DB to rectangle DB by BG is like the ratio of AC to BG: so the ratio of EC to FG is less than the ratio of AC to GB. So line FE will cut off a ratio of EC to FG less than all the ratios that the lines drawn from point H cut off, while cutting all along line GD.

And this problem will be synthesized thus: Let the rest be as before. Let line GD be cut in half at point F, and let FH be joined and produced in a straight line to point E. So the line FE cuts off a ratio of EC to GF less than all the ratios that the lines drawn from point H cut off, while cutting all along line GD.

So, if the ratio being given in the synthesis is the same as the ratio of EC to FG, line EF alone will do what the problem requires. And if the ratio is less, then the problem will not be synthesized. But if the ratio is greater, then the problem will be synthesized.

For let the given ratio be the ratio of J to O, which is greater than the ratio of EC to FG. And let it be made that, as the ratio of J to O, so the ratio of rectangle CH by HD, the equal of rectangle EC by DF, to rectangle P. However the ratio of J to O is greater than the ratio of EC to FG. So the ratio of rectangle EC by DF to rectangle P is greater than the ratio of EC to FG. However the ratio of EC to FG is like the ratio of rectangle EC by DF to rectangle DF by FG. So the ratio of rectangle EC by DF to rectangle P is greater than the ratio of rectangle EC by DF to rectangle DF by FG. So rectangle P is less than rectangle DF by FG. And thus it is possible to apply to line DG a rectangle equal to rectangle P and deficient by a square, and that application will obtain in two ways. So let the two points of application be the two points B, K. And let the two lines BH, KH be joined and produced to points A, L. Then I say that each of the two lines AB, KL will do what the problem requires.

For, since the ratio of J to O is like the ratio of rectangle CH by HD to rectangle P, while rectangle CH by HD is equal to rectangle LC by DK, and rectangle P is equal to rectangle DK by KG, then the ratio of J to O is like the ratio of rectangle LC by DK to rectangle DK by KG, which is the same as the ratio of LC to KG. So line KL will do what the problem requires. And the same argument also makes it clear that line AB will make the problem work. So it has become clear that the problem will be synthesized in two ways. And that is what we wanted to show.

Case 3. And let a line HB be drawn according to the third case, cutting off a recipient ratio of CA to BG. So the ratio of rectangle CA by BD to rectangle BG by BD is a recipient. However rectangle BD by CA is a recipient. [So rectangle BG by BD is a recipient], and it has been applied to line GD with an excess of the square on DB. So point B is a recipient. But point H is a recipient. So line HB is positioned. And the ratio

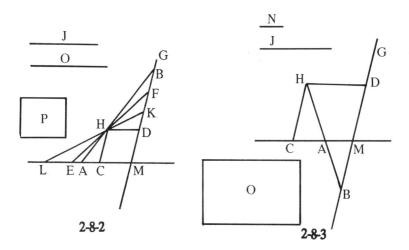

2-8-2

2-8-3

being given in the synthesis will have to be less than the ratio of CM to MG.

For, since line MC is greater than line CA while MG is less than BG, the ratio of MC to CA will be greater than the ratio of MG to BG. And by alternation, the ratio of CM to MG will be greater than the ratio of CA to BG. However the ratio of CA to BG is a recipient. So in the synthesis it will be necessary that the given ratio be less than the ratio of CM to MG.

And this problem will be synthesized thus: With the rest as before, let the given ratio be the ratio of N to J, which is less than the ratio of CM to MG. And let it be made that, as the ratio of N to J, so the ratio of rectangle CH by HD, the equal of rectangle CM by MD, to rectangle O. Now the ratio of N to J is less than the ratio of CM to MG. So the ratio of rectangle CM by MD to rectangle O is less than the ratio of CM to MG. However the ratio of CM to MG is like the ratio of rectangle CM by MD to rectangle MD by MG. So the ratio of rectangle CM by MD to rectangle O is less than the ratio of rectangle CM by MD to rectangle GM by MD. So rectangle O is greater than rectangle GM by MD. So, if we want to apply to line GD a rectangle equal to rectangle O with the excess of a square, then it will pass beyond point M. So let rectangle O be equal to rectangle GB by BD, and let HB be joined. Then I say that HB will do what the problem requires.

For, since the ratio of N to J is like the ratio of rectangle CH by HD to rectangle O, while we made rectangle O equal to rectangle GB by BD, so the ratio of N to J is like the ratio of CH by HD to rectangle GB by BD. However rectangle CH by HD is equal to rectangle BD by AC. So rectangle N to J is like the ratio of rectangle CA by BD to rectangle GB by BD. And the ratio of rectangle CA by BD to rectangle GB by BD is like the ratio of CA to BG. So the ratio of N to J is like the ratio of CA to BG. So line HB will do what the problem requires, and that is what we wanted to show.

Case 4. And let a line HA be drawn according to the fourth case, cutting off a recipient ratio of CA to BG. So the ratio of rectangle CA by BD to rectangle GB by BD is a recipient. However rectangle CA by BD is equal to rectangle CH by HD. So rectangle GB by BD is recipient, and it has been applied to line DG with an excess of the square on BD. So line BD is recipient. But point H is recipient. So line HB is positioned.

And it will be necessary that the ratio being given in the synthesis be greater than the ratio of CM to MG. For, since line CA is longer than line CM while BG is less than GM, the ratio of CA to CM is greater than the ratio of BG to GM. And by alternation, the ratio of AC to BG is greater than the ratio of CM to GM. However the ratio of AC to BG is a recipient. So it will be clear that the ratio being given in the synthesis will have to be greater than the ratio of CM to MG.

And this problem will be synthesized thus: With the rest as before, let the given ratio be the ratio of N to J, which is greater than the ratio of CM to MG. And let it be made that, as the ratio of N to J, so the ratio of rectangle CH by HD, the equal of rectangle CM by MD, to rectangle O. So, since the ratio of N to J is greater than the ratio of CM to MG, while the ratio of N to J is like the ratio of rectangle CM by MD to rectangle O, and the ratio of CM to MG is like the ratio of rectangle CM by MD to rectangle DM by MG, so the ratio of

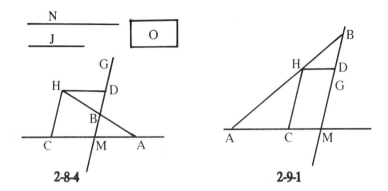

2-8-4   2-9-1

rectangle CM by MD to rectangle O is greater than the ratio of
rectangle CM by MD to rectangle GM by MD. So rectangle O is
less than rectangle GM by MD. So, if we want to apply to line
GD a rectangle equal to rectangle O, which is excessive by a
square, then the point of application will fall within point M. So
let rectangle O be the equal of rectangle GB by BD, and let line
HB be joined and produced to A. Then I say that line HA will
do what the problem requires.

For, since the ratio of N to J is like the ratio of rectangle
CH by HD, the equal of rectangle CA by BD to rectangle GB
by BD, while the ratio of rectangle CA by BD to rectangle GB
by BD is like the ratio of CA to BG, so the ratio of N to J is
like the ratio of CA to BG. And that is what we wanted to
show.

Locus 9. And also let one of the two parallels fall above point G
as line HD. Accordingly, it is clear that the cases, that is, the
ways in which the lines drawn from H fall, will be in four
positions.

Case 1. So let a line BA be drawn according to the first case,
cutting off a recipient ratio of AC to BG. So the ratio of
rectangle AG by BD to rectangle BG by BD is a recipient.
However rectangle AC by DB is a recipient. So rectangle GB by
BD is a recipient. But it has been applied to line GD with an

excess of the square on BD. So point B is a recipient. And point H is a recipient. So line HA is positioned. And the synthesis of the problem will be clear by its similarity to what has been previously stated. And that is what we wanted to show.

Case 2. And let a line HB be drawn according to the second case, cutting off a recipient ratio of CA to BG. So the ratio of rectangle AC by DB to rectangle DB by BG is a recipient. However rectangle BD by CA is a recipient. So rectangle DB by BG is a recipient. But it has been applied to line DG with the excess of a square. So point B is recipient. But point H is recipient. So line HB is positioned.

And it will be necessary that the ratio being given in the synthesis be less than the ratio of CM to MG. For, since line MC is longer than line CA while line MG is less than line GB, the ratio of MC to CA is greater than the ratio of MG to GB. And likewise by alternation. However the ratio of CA to BG is a recipient. So it will be necessary that the ratio being given in the synthesis be less than the ratio of CM to MG.

And the problem will be synthesized thus: Let the rest be the same as before, only let the given ratio be the ratio of N to J, which is less than the ratio of CM to MG. And let it be made that, as the ratio of N to J, so the ratio of rectangle CH by HD, the equal of rectangle CM by MD, to rectangle O. But the ratio of N to J is less than the ratio of CM to MG. So the ratio of rectangle CM by MD to rectangle O is less than the ratio of CM to MG. However the ratio of CM to GM is like the ratio of rectangle CM by MD to rectangle DM by MG. So the ratio of rectangle CM by MD to rectangle O is less than the ratio of rectangle CM by MD to rectangle DM by MG. So let there be applied to line DG a rectangle equal to rectangle O with the excess of a square, namely, rectangle DB by BG. Then it will be clear that line HB will do what the problem requires, and that is what we wanted to show.

Case 3. And let a line HA be drawn according to the third case, cutting off a recipient ratio of BG to CA. So the ratio of

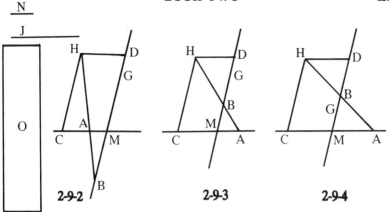

2-9-2   2-9-3   2-9-4

rectangle BG by BD to rectangle CA by BD is a recipient.
However the rectangle CA by BD is recipient. So rectangle GB
by BD is recipient. But it has been applied to line DG with the
excess of a square. So point B is recipient. But point H is
recipient. So line HA is positioned.

And it will be necessary that the ratio being given in the
synthesis be greater than the ratio of CM to MG. For, since the
ratio of AC to CM is greater than the ratio of BG to GM, by
alternation it will be likewise. But the ratio of CA to BG also
is a recipient. So it has become clear that the ratio being given
in the synthesis must be greater than the ratio of CM to MG.

And since what we have stated has become clear, its
synthesis is clear accordingly. And that is what we wanted to
show.

Case 4. And let a line HA be drawn according to the fourth
case, cutting off a recipient ratio of CA to BG. So the ratio of
rectangle CA by BD to rectangle BG by BD is a recipient.
However rectangle CA by DB is recipient. So rectangle DB by
BG is recipient. But it has been applied to line GD with the
defect of a square. So point B is recipient. But point H is
recipient. So line HA is positioned. And that is what we wanted
to show.

And we will find that as follows: With the rest the same as
before, let GD be cut in half at point B. And let HB be joined
and produced in a straight line to point A. Then I say that line

HA will cut off a ratio of CA to BG, which is less than all the ratios that the lines drawn from point H cut off, while cutting all along line GD.

For let line HK be drawn. Then, since line GB is equal to line BD, rectangle GB by BD will be greater than rectangle GE by ED. However rectangle CA by DB is equal to rectangle CK by ED, because each of them is equal to rectangle CH by HD. So the ratio of rectangle CA by BD to rectangle GB by BD is less than the ratio of rectangle CK by ED to rectangle GE by ED. However the ratio of rectangle CA by BD to rectangle GB by BD is like the ratio of CA to GB, while the ratio of rectangle CK by ED to GE by ED is like the ratio of CK to GE. So the ratio of CA to GB is less than the ratio of CK to EG. So line HA will cut off a ratio of AC to GB less than all the ratios that the lines drawn from point H cut off, while cutting all along GD.

And since the problem has been limited in this way, so from what was previously stated it will necessarily be synthesized in two ways, on the two sides of line HA, cutting GB, BD. And that is what we wanted to show.

Locus 10. And now let the two lines drawn through point H parallel to the lines AM, MK fall above the two points G, C as the lines EH, HD. So it is clear that the lines that are drawn through point H will fall in five positions.

Case 1. So let a line AK be drawn according to the first case, cutting off a recipient ratio of KG to AC. Let line HG be joined and produced in a straight line to point F. And let a line FB be drawn through point F parallel to line MK. And let HAK be produced to point B. So line FB is positioned. Then, since the ratio of GK to AC is a recipient, while the ratio of GK to FB is a recipient, then the ratio of FB to AC is a recipient. So, since the two lines AF, FB are positioned, and the place which is passed by on line AF is at point C and the place which is passed by on line FB is at point F, and the recipient point H is within angle AFB, and the parallel line HE which passes

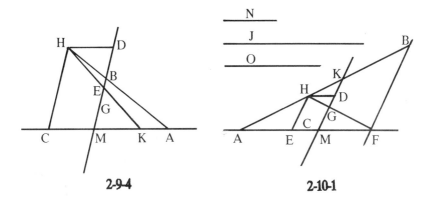

2-9-4   2-10-1

through point H falls above C, and line AB was drawn
according to the first case cutting off a recipient ratio of FB to
CA: so line AB is positioned, because it resembles the first case
of the sixth locus, which has no limitation. And the synthesis of
this problem is clear.

This problem will be synthesized thus: With the rest as
before, let the given ratio be the ratio of N to O. And let it be
made that, as the ratio of GH to HF, so the ratio of N to J. And
let a line AB be drawn according to the first case of the sixth
locus, cutting off a ratio of FB to AC the same as the ratio of J
to O. So it is clear that line AB will do what the problem
requires, and that is what we wanted to show.

Case 2. And let a line HK be drawn according to the second
case, cutting off a recipient ratio of KG to AC. Let line HG be
produced to F. And let a line be drawn through point F parallel
to KM, meeting HK at point B. And the ratio of AC to KG is a
recipient. However the ratio of KG to BF is a recipient. So the
ratio of BF to AC is a recipient. So line HB is a recipient, since
it resembles the second case of the sixth locus, which has a
limitation.

And we will find that as follows: With the rest as before, let
the line FA be a mean proportional between the two lines EF,
FC. And let HA be joined and produced in a straight line to
point B. Then I say that line HB will cut off a ratio of KG to
CA less than all the ratios that the lines drawn from point H

cut off, while cutting all along line CE.

For let another line be drawn, such as line HP. So, since line FA is a mean proportional between lines EF, FC and the two lines PF, EF are positioned, and the places which are passed by on them are point F on line PF and point C on line EF, and the line HE parallel to line FP falls above point C: therefore, because of what has been shown in the second case of the sixth locus, the ratio of FB to AC is least. So the ratio of BF to CA is less than the ratio of PF to CJ. And by alternation, the ratio of BF to FP is less than the ratio of AC to CJ. However the ratio of BF to FP is like the ratio of KG to GO. So the ratio of KG to GO is less than the ratio of AC to CJ. And by alternation, the line HB will cut off a ratio of KG to CA which is less than all the ratios that the lines drawn from point H cut off.

And the problem will be synthesized thus: With the rest as before, let there be taken a mean proportional between lines EF & FC, namely, line FA. And let HA be joined and produced in a straight line to point B. So line HK will cut off a ratio of KG to CA less than all the ratios that the lines drawn from point H cut off, while cutting all along line CE.

And when the ratio being given in the synthesis is the same as the ratio of KG to CA, line HK will likewise do what the problem requires. And if it is less than it, then the problem will not be synthesized. But if it is greater than it, then it will be clear from what has been previously stated that the problem will obtain in two ways, on both sides of line HK, cutting EA, AC.

So let the given ratio be the ratio of R to N, which is greater than the ratio of KG to CA, and let it be made that, as the ratio of GH to FH, so the ratio of R to X. So the ratio of R to X will be like the ratio of KG to FB. And the ratio of R to N is greater than the ratio of KG to CA. So, by way of equality, the ratio of X to N will be greater than the ratio of FB to CA. However the ratio of BF to CA is the least of the ratios, according to what has been made clear in the second case of the sixth locus. So two lines will be drawn, on both sides of line HB. So let line HP be drawn, cutting off the ratio of PF to CJ

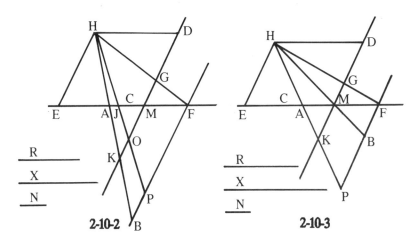

2-10-2   2-10-3

the same as the ratio of X to N. Then it will be clear that line
HP will do what the problem requires. And likewise also it can
be shown that the other line will do so too. And that is what we
wanted to show.

Case 3. And let a line HK be drawn according to the third case,
cutting off a recipient ratio of KG to CA, and let it be
produced to point P. However the ratio of KG to PF is a
recipient. So the ratio of PF to CA is a recipient. So line HP is
positioned, because it resembles the third case of the sixth
locus.

And we will find that as follows: With the rest as before, let
HM be joined and produced to point B. So it is clear that the
ratio of GM to MC is less than the ratio of KG to CA. So the
ratio being given in the synthesis will have to be greater than
the ratio of GM to MC.

And this problem will be synthesized thus: With the rest
the same, let the given ratio be the ratio of R to N, which is
greater than the ratio of GM to MC. And let it be made that, as
the ratio of MH to HB, so the ratio of R to X. So it is clear, by
way of equality, that the ratio of BF to CM will be less than the
ratio of X to N. So we want to make a line pass through point
H, cutting off from the two lines MC, FP two lines whose ratio
to one another is like the ratio of X to N. So it is clear that it
will cut line CM, as it is obvious that lines near point F always

cut off ratios less than the ratios that lines distant from it cut off. And let it be drawn as line HP, cutting off a ratio of PF to CA the same as the ratio of X to N. So it is clear that line HP will do what the problem requires. And that is what we wanted to show.

Case 4. And let a line HP be drawn according to the fourth case, cutting off a recipient ratio of AG to BC. However the ratio of AG to PF is a recipient. So the ratio of PF to CB is a recipient. And line HP is positioned, because it resembles that third case cited in the previous case.

And we will find that as follows: With the rest the same, let HM be joined and produced to point E. So it will be clear that the ratio being given in the synthesis will have to be less than the ratio of GM to MC.

And the problem will be synthesized thus: With the rest the same as before, let the given ratio be the ratio of R to N, which is less than the ratio of GM to MC. And let it be made that, as the ratio of GH to HF, so the ratio of R to X. So it is clear, by way of equality, that the ratio of FE to MC will be greater than the ratio of X to N. And if we want to pass a line through point H, cutting off segments from the two lines CF, FE whose ratio to one another is like the ratio of X to N, it will be clear that that line will cut line MF, since it is clear that lines near point F cut off ratios greater than the ratios that lines distant from it cut off. Let that line be drawn. However line HP will cut off a ratio of PF to BC the same as the ratio of X to N. So it is clear that line HP will do what the problem requires. And that is what we wanted to show.

Case 5. And let a line AB be drawn according to the fifth case, cutting off a recipient ratio of AG to CB. However the ratio of GA to FO is a recipient. So the ratio of FO to BC is a recipient. So line HB is positioned, because it resembles the fourth case of the sixth locus which <has> a limitation.

And we will find the problem thus: With the rest as before, let there be taken a mean proportional FB between lines EF, FC. And let line HB be joined. So line HB will cut off the ratio

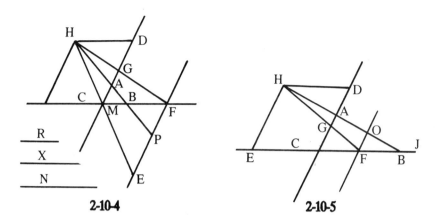

2-10-4   2-10-5

of FO to BC which is greater than all the ratios that the lines
drawn from point H cut off, while cutting all along line FJ. So
it is clear that line HB will cut off a ratio of GA to BC greater
than all the ratios that the lines cut off, while cutting line GD.

And this problem will be synthesized thus: With the rest as
before, let there be taken a mean proportional FB between lines
EF, FC. And let HB be joined. So line HB will cut off a ratio of
GA to BC which is greater than all the ratios that the lines
drawn from point H cut off, while cutting all along line GD. So
it will be clear from what has been previously stated that the
problem will obtain in two ways, on both sides of line HB,
cutting AG, AD. And that is what we wanted to show.

**And let the lines HE, HD which are drawn from point H
parallel to the two lines MG, CM fall lower than the two points
C, G. And let point G be joined to point H and produced in a
straight line. Then it will either fall on point C or to one side
of it.**

Locus 11. So first let it fall on it. Then it is clear that the lines
drawn from point H fall in four ways.

Case 1. So let a line AB be drawn according to the first case,
cutting off a ratio of BG to CA. And let a line CF be drawn

through point C parallel to line MG. So, since the ratio of BG to CA is a recipient, while the ratio of BG to CF is a recipient, because it is the same as the ratio of GH to CH, then the ratio of CF to CA is a recipient. So, since there are two positioned lines, namely, lines AM & CF, and the place which is passed by on each of them is at point C, and the recipient point H is within angle FCM, and line HA has been drawn cutting off a recipient ratio of FC to CA: then line AB is positioned, because it resembles the first case of the third locus, which <has> a limitation.

And in the synthesis it will be necessary that the given ratio be less than the ratio of DG to EC. For let line DH be produced to point K. Now since the ratio of HE to EC is greater than the ratio of HE to EA, while the ratio of HE to EA is like the ratio of FC to CA, so the ratio of HE to EC is greater than the ratio of FC to CA. But line EH is equal to line CK. So the ratio of line CK to line CE is greater than the ratio of FC to CA. So, by alternation, the ratio of CK to CF is greater than the ratio of EC to CA. However the ratio of KC to CF is like the ratio of DG to GB. So the ratio of DG to GB is greater than the ratio of EC to CA. And by alternation, it still is. And the ratio of GB to CA is a recipient. So it will be necessary that the given ratio be less than the ratio of DG to EC.

And this problem will be synthesized thus: With the rest as before, let the given ratio be the ratio of P to N, which is less than the ratio of DG to EC. And let it be made that, as the ratio of CH to GH, so the ratio of X to P. However the ratio of CH to GH is like the ratio of CK to DG, which is the same as the ratio of EH to DG. So the ratio of X to P is like the ratio of HE to GD. But the ratio of P to N is less than the ratio of DG to EC. So, by way of equality, the ratio of X to N will be less than the ratio of EH to EC. So, if we make the ratio of X to N like the ratio of HE to another line, then that line will be greater than EC. So let it have that ratio to line EA. And let HA be joined and produced in a straight line. So it is clear that it will do what the problem requires. And that is what we wanted to show.

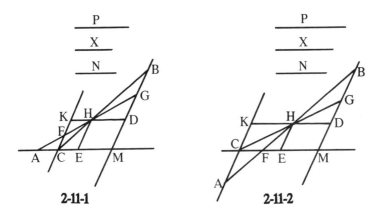

2-11-1   **2-11-2**

Case 2. And let a line FB be drawn according to the second case, cutting off a recipient ratio of GB to CF. And let it be produced to point A. However the ratio of GB to AC is a recipient. So the ratio of AC to CF is a recipient. And line BA is positioned, because it resembles the second case of the third locus in which there <is> a limitation.

And it will be necessary that in the synthesis the given ratio be greater than the ratio of GD to EC. For let DH be produced to point K. Then, since the ratio of HE to EF, that is, the ratio of AC to CF, is greater than the ratio of EH to EC, that is, the ratio of KC to CE, and by alternation, the ratio of AC to CK is greater than the ratio of FC to CE, while the ratio of AC to KC is like the ratio of BG to GD, so the ratio of BG to GD is greater than the ratio of FC to CE. And by alternation, the ratio of BG to FC is greater than the ratio of GD to EC. And the ratio of BG to FC is a recipient. So it will be necessary that the given ratio be greater than the ratio of GD to EC.

And the problem will be synthesized thus: With the rest the same as before, let the given ratio be the ratio of P to N, which is greater than the ratio of GD to CE, and let it be made that, as the ratio of CH to HG, so the ratio of X to P. But the ratio of P to N is greater than the ratio of GD to EC. So, by way of equality, the ratio of X to N is greater than the ratio of CK to EC, that is, than the ratio of HE to EC. So, since there are two positioned lines EC & AK, and the place which is passed by on them both is at point C, and the given ratio is greater than the ratio of HE to EC, so we can cut off a ratio the same as the ratio of X to N, while the line from point H which cuts them off cuts CE. So let it be line AB. So it is clear that line AB will do what the problem requires. And that is what we wanted to show.

Case 3. And let a line HA be drawn according to the third case, cutting off a recipient ratio of AG to OC, and let it be produced to point B. However the ratio of AG to CB is a recipient. So the ratio of BC to OC is a recipient. And line AB is positioned, because it resembles the third case of the third locus.

And we will find that in this way: With the rest as before, it will be necessary that the ratio being given in the synthesis be greater than the ratio of GM to MC, because the ratio of AG to OC is greater than the ratio of GM to MC, as we have premised.

And the problem will be synthesized in this way: With the rest as before, let HM be joined and produced to point K. Let the given ratio be the ratio of N to P, which is greater than the ratio of GM to MC. And let it be made that, as the ratio of GH to CH, so the ratio of N to J. However the ratio of GH to CH is like the ratio of MG to KC. So the ratio of N to J is like the ratio of MG to KC. And the ratio of N to P is greater than the ratio of GM to MC. So it is clear, by way of equality, that the ratio of J to P will be greater than the ratio of CK to CM. So let a line AB be drawn cutting off a ratio of BC to CO, which is the same as the ratio of J to P. So it will also cut line EM, because lines near point C will always cut off ratios greater than the ratios that lines distant from it cut off, as we can make clear. So it is clear that line HA does what the problem requires. And that is what we wanted to show.

Case 4. And let a line AHB be drawn according to the fourth case, cutting off a recipient ratio of OG to AC. However the ratio of OG to BC is a recipient. So the ratio of BC to CA is a recipient. And line AB is positioned. And it will be necessary

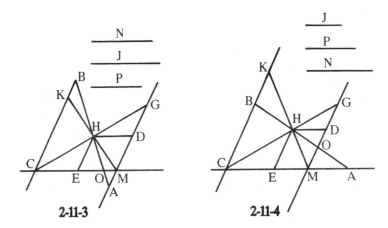

2-11-3   2-11-4

that the ratio being given in the synthesis be less than the ratio of GM to MC, as we have already explained.

And this problem will be synthesized in this way: With the rest the same, let the given ratio be the ratio of J to N, which is less than the ratio of GM to MC, and let HM be joined and produced to point K. And let it be made that, as the ratio of GH to CH, so the ratio of J to P. However the ratio of GH to HC is like the ratio of MG to KC, while the ratio of MG to MC is greater than the ratio of J to N. So it is clear, by way of equality, that the ratio of KC to CM will be greater than the ratio of P to N. And let a line HA be drawn, cutting off a ratio of BC to CA the same as the ratio of P to N. So AH ineluctably will cut off line MA, because lines near point C will always cut off ratios greater than the ratios that lines distant from it cut off. So it is clear that line AB will do what the problem requires. And that is what we wanted to show.

Locus 12. And also let the line that is drawn from point H to point G fall above point C as line GF. Then it is clear that the lines drawn from point H will fall in five ways.

Case 1. So let a line HB be drawn according to the first case, cutting off a recipient ratio KG to BC. And let a line FA be drawn through point F parallel to line MG. However the ratio of KG to FA is a recipient, because it is the same as the ratio of GH to HF. So the ratio of FA to BC is a recipient. So, since two lines BF, FA are positioned, and the place which is passed by on line FA is at point F and the place which is passed by on line FB is at point C, and the recipient point H is within angle AFM, and the parallel line that passes through point H, namely, line HE, falls above point C, and line BA is drawn cutting off a recipient ratio of AF to CB: then line AB will be positioned, because it resembles the first case of the sixth locus.

And clearly it will be necessary that the ratio being given in the synthesis be less than the ratio of GM to MC, because line KG is less than line GM while line BC is greater than line CM.

And this problem will be synthesized in this way: With the rest as before, let the given ratio be the ratio of N to O, which is less than the ratio of GM to MC. And let MH be joined and produced in a straight line to point L. And let it be made that, as the ratio of GH to HF, that is, the ratio of GM to LF, so the ratio of N to J. So the ratio of N to J is like the ratio of GM to FL. But the ratio of N to O is less than the ratio of GM to MC. So, by way of equality, the ratio of J to O will be less than the ratio of LF to CM. And by inversion, the ratio of O to J will be greater than the ratio of CM to FL. So, if we make the ratio of O to J like the ratio of MC to another line, then that line will be less than FL. So let it have that ratio to line FA. And let AH be joined and produced in a straight line. And likewise it will be clear that, if according to the first case of the sixth locus, we wanted to draw through point H a line cutting off from the two lines FL, FB two segments whose ratio to one another is like the ratio of J to O, then it is clear that it will cut line MB, because lines near point C will always cut off ratios greater than the ratios that lines near it cut off. So let a line AB be drawn, cutting off a ratio of AF to CB, which is the same as the ratio J to O. Then it is clear that line AB will do what the problem requires. And that is what we wanted to show.

Case 2. And let a line HB be drawn according to the second case, cutting off a recipient ratio of GB to KC. However the ratio of GB to AF is a recipient. So the ratio of AF to KC is a recipient. So line AB is positioned, because it also resembles the

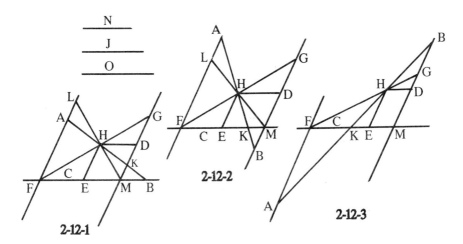

2-12-1

2-12-2

2-12-3

first case. And clearly it will be necessary that the ratio being given in the synthesis be greater than the ratio of GM to MC.

And this problem will be synthesized in this way: With the rest the same as before, let HM be joined and produced in a straight line to point L. And the rest of what remains is like what was said in the preceding <case>. And that is what we wanted to show.

Case 3. And let a line HB be drawn according to the third case, cutting off a recipient ratio of GB to KC. And let line HK be produced to point A. However the ratio of GB to AF is a recipient. But the ratio of GB to KC is a recipient. So the ratio of AF to KC is a recipient, and line AB is positioned, because it resembles the second case of the sixth locus, which has a limitation.

And we will find that in this way: With the rest as before, let there be taken a mean proportional between lines EF & FC, namely, line FK. Then let HK be joined and produced in a straight line to points A, B. So line HA will cut off a ratio of FA to CK less than the ratios that the lines drawn from point H cut off, while cutting all along line EC. And it will also be clear that line BK will cut off a ratio of GB to KC which is less than all the ratios that the lines drawn from point H cut off, while cutting all along line EC.

So, since we have delimited that, the synthesis of this problem will be clear, and it will obtain in two ways, on both sides of line BK, while cutting EK, KC. And that is what we wanted to show.

Case 4. And let a line AB be drawn according to the fourth case, cutting off a recipient ratio of GB to KC. Also let a parallel line be drawn which passes through point F. So the ratio of GB to FA is a recipient. However the ratio of GB to KC is a recipient. So line AB is positioned, because it resembles the third case of the sixth locus, which has no limit.

And the synthesis of that is clear similarly to what we have said in the preceding case. And that is what we wanted to show.

Case 5. And let a line AB be drawn according to the fifth case, cutting off a recipient ratio of BG to AC. However the ratio of GB to FK is a recipient. So the ratio of FK to AC is a recipient. And line AB is positioned, because it resembles the fourth case of the sixth locus, which has a limitation.

And we will find it in this way: With things the same as before, let there be taken a mean proportional between lines EF & FC, namely, line FA. And let HA be joined. So line HA will cut off a ratio of FK to CA, which is greater than all the ratios that the lines drawn from point H cut off, while cutting all along line CA. And it is clear that line AB will cut off a ratio of BG to AC which is greater than all the ratios that the lines drawn from point H cut off, while cutting all along line CJ.

And since what we have stated has become clear, the synthesis of this problem is clear. And it will obtain in two ways, on both sides of AB, cutting AF, AJ. And that is what we wanted to show.

Locus 13. And also let the line drawn from point H to point G and produced in a straight line fall below point C as line FG. Then it is clear that the lines that are drawn from point H will cut off in five ways.

Case 1, 2 & 3. Let AB be drawn according to the first, second and third cases, cutting off a ratio of GB to CA, while a line

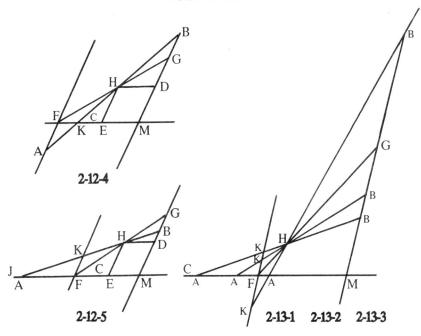

2-12-4

2-12-5 2-13-1 2-13-2 2-13-3

FK is drawn through point F parallel to line MG. So, since the ratio of BG to AC is a recipient and the ratio of BG to FK is a recipient, the ratio of KF to CA is a recipient. And since there are two positioned lines in a plane, namely, FK & AM, and the place which is passed by on line FK is at point F, and the place which is passed by on line MA is at point C, and the recipient point H is within angle KFM, and line AB was drawn cutting off a recipient ratio of FK to CA: so line AB is positioned, because in the first case it resembles the first case of the fourth locus, and in the second the second case, and the third the third, and its ratio is not determined. And the synthesis of these will be clear according to what we have discussed. And that is what we wanted to show.

Case 4 And let a line HB be drawn according to the fourth case, cutting off a recipient ratio of UU to CK, while HB is produced to point A. However the ratio of BG to FA is a recipient. So the ratio of AF to KC is a recipient. So line AB is positioned, because it resembles the fourth case of the fourth locus. And clearly it will be necessary that the ratio being given in the synthesis be greater than the ratio of GM to MC.

And the synthesis of this problem will be like this: With things the same as before, let the given ratio be the ratio of J to P, which is greater than the ratio of GM to MC. Let MH be joined and produced in a straight line to point L. And let it be made that, as the ratio of GH to HF, so the ratio of J to O. So it will be clear, by way of equality, that the ratio of LF to CM will be less than the ratio of O to P. And let AB be drawn through point H cutting off a ratio of FA to CK the same as the ratio of O to P. It will cut line EM, since lines near point C will ineluctably cut off ratios greater than the ratios that lines distant from it cut off, as was previously explained. So it will be clear that line AB will do what the problem requires. And that is what we wanted to prove.

Case 5. And let a line AB be drawn according to the fifth case, cutting off a recipient ratio of GK to BC. However the ratio of GK to AF is a recipient. So the ratio of AF to BC is a recipient. So line AB is positioned, because it also resembles the fourth case. And it is clear that the ratio being given in the synthesis will necessarily be less than the ratio of GM to MC. So, since what we have stated will be clear in accordance with the cited example, so its synthesis will also be clear by following the example of the previously discussed case. And that is what we wanted to show.

Locus 14. And also let one of the two lines drawn parallel to the lines AF, GM fall below point C, namely, HE, while the other falls above point G, namely, HD. Then it will be clear that the lines drawn from point H will fall in five positions.

Case 1. So let a line AB be drawn according to the first case, cutting off a recipient ratio of GB to AC. Let HG be joined and and produced to point F. And let a line FK be drawn through point F parallel to line MB. And let AB be produced to point K. However the ratio of GB to FK is a recipient. So the ratio of FK to AC is a recipient. So, since there are two

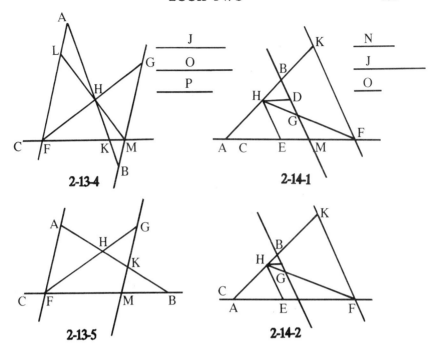

2-13-4

2-14-1

2-13-5

2-14-2

positioned lines, namely, the two lines AF & FK, and the place
which is passed by on FK is at point F, and the place which is
passed by on AF is at point C, and the recipient point H is
within angle AFK, and the line which is parallel to line FK and
passes through point H, namely, line HE, falls below point C,
and line AK has been drawn cutting off a recipient ratio of FK
to AC: so line AK is positioned, because it resembles the first
case of the seventh locus, which has no limitation.

And the problem will be synthesized thus: With the rest as
before, let the given ratio be the ratio of N to O. And let it be
made that, as the ratio of GH to FH, so the ratio of N to J. And
let line AB be drawn according to the first case of the seventh
locus, cutting off a ratio of KF to AC the same as the ratio of J
to O. So it will be clear that line AK will do what the problem
requires. And that is what we wanted to show.

Case 2. And let a line AB be drawn according to the second
case, cutting off a recipient ratio of GB to AC. Let line AB be
produced to point K. However the ratio of GB to FK is a

recipient. So the ratio FK to AC is a recipient. So line AK is positioned, because it resembles the second case of the seventh locus, which has a limitation.

We will find it in this way: With the rest as it was, let there be taken a mean proportional line FA between lines FC, FE. Let AB be joined and produced to point K. So line AK will cut off a ratio of FK to AC, which is less than all the ratios that the lines drawn from point H cut off, while cutting all along line EC. So it will be clear from the limitations previously stated that line AK will cut off a ratio of GB to AC, which is less than all the ratios that the lines drawn from point H cut off, while cutting all along line EC.

And from that it will also be clear how the synthesis of these problems will obtain. That is, that they will be synthesized in two ways, on both sides of AK, cutting CA, AE. And that is what we wanted to show.

Case 3. And let a line HB be drawn according to the third case, cutting off a recipient ratio of GB to CA. And let it be produced to point K. However the ratio of GB to KF is a recipient. So the ratio of KF to CA is a recipient. So line HK is positioned, because it resembles the third case of the seventh locus. And it is clear that the ratio being given in the synthesis will necessarily be greater than the ratio of GM to MC.

And this problem will be synthesized thus: With the rest as before, let the given ratio be the ratio of N to O, which is greater than the ratio of GM to MC. Let HM be joined and produced to point L. And let it be made that, as the ratio of GH to HF, so the ratio of N to J. So it is clear, by way of equality, that the ratio of LF to CM will be less than the ratio of J to O. And if HK is drawn cutting off a ratio of KF to CA the same as the ratio of J to O, it ineluctably cuts line EM, because it is clear that lines near the point C will always cut off ratios greater than the ratios that lines distant from it cut off. So line KH will do what the problem requires, and that is what we wanted to show.

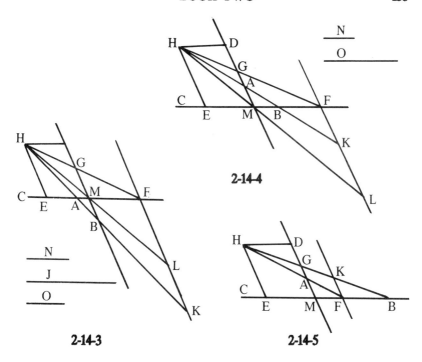

2-14-4

2-14-3   2-14-5

Case 4. And let a line HB be drawn according to the fourth case, cutting off a recipient ratio of AG to BC, and let it be produced to point K. However the ratio of AG to KF is a recipient. So the ratio of KF to BC is a recipient. So line HK is positioned, because it resembles the case which went before. And clearly the ratio being given in the synthesis will necessarily be less than the ratio of GM to MC.

And the synthesis of this problem will be thus: With the rest as before, let the given ratio be the ratio of N to O, which is less than the ratio of GM to MC. Let HM be joined and produced to point L. And the rest will be clear from the past figure and from what has been previously stated. And that is what we wanted to show.

Case 5. And let a line HB be drawn according to the fifth case, cutting off a recipient ratio of AG to BC. However the ratio of GA to FK is a recipient. So the ratio of FK to CB is a recipient. So line HB is positioned, because it resembles the fourth case of the fifth locus, which has a limitation.

And we will find this problem thus: Let the rest be as before. And let line FB be made a mean proportional between lines CF, FE. And let HB be joined. Then line HB will cut off the ratio of FK to BC, which is greater than all the ratios that the lines drawn from point H cut off, while cutting all along line BF. And it will also be clear from the limitations previously stated that line HB will cut off a ratio of GA to BC, which is greater than all the ratios that the lines drawn from point H cut off, while cutting all along line GD.

And since what we have stated about that has become clear, it will become clear that the synthesis of this problem will obtain in two ways, on both sides of line HB, cutting GA, AD. And clearly all of that [will be easy to find] from what we have previously stated. And that is what we wanted to show.

TRANSLATOR'S REMARKS
ON THE METHODS USED IN THIS TRANSLATION

DESCRIPTION OF MANUSCRIPTS USED

1.) Ayasofya 4830 (MS A), microfilm supplied through courtesy of the Embassy of Turkey. This text is missing an extensive section from near the end of Book 1, Locus 4, Case 3 to near the end of Book 1, Locus 6, Case 2. It is also lacking the last page of the summary at the end of Book 2.

Though more difficult to read than the Bodleian MS, the text given is freer from dittography, haplography, and other copyist's errors.

2.) Bodleian Arch. Seld. A 32 (MS B), microfilm supplied through the Keeper of Manuscripts of the Bodleian Library. This text served as the basis of Halley's 1706 Latin paraphrase. This MS is substantially complete, except for haplography.

Together, the two MSS make up a substantially complete text.

HOW THESE MANUSCRIPTS WERE USED

1.) To distinguish Loci and Cases.

2.) To establish general orientation and lettering of figures.

3.) To determine number of figures for an argument. It makes considerable difference in understanding the relation between the analysis, diorismos, and synthesis. For example, it is only in the synthesis figure that the given ratio is first displayed.

4.) To determine order of mathematical steps in the arguments.

5.) To discern the logical argumentation. For example, much is revealed by where the definite article occurs in the analysis.

6.) As a guide to divining the original Greek terminology. For example, we conjecture that the Greek had ἐστὶ δεδομένον where the Arabic has 'is known', while the Greek had ἐστὶ δοθέν where the Arabic has 'is assigned'; and on the rare occasion where the Arabic has 'is given', the Greek probably had δέδοται.

DEPARTURES FROM A LITERAL TRANSLATION

1.) In conformity with Greek mathematical usage, we have changed most of personal verb forms of the Arabic to impersonals. We have retained just a sprinkling of personals, as Apollonius himself seems to have done elsewhere.

2.) The construction of a fourth proportional in the manuscripts usually seems to compare the ratios incorrectly. We have changed this as was required by the sense.

3.) The figures have been reoriented somewhat, although we tried to keep the overall pattern. Many were rotated 180 degrees. This seemed to bring the figures into conformity with internal directional references in the text.

4.) We have interchanged the figures for Cases 2 & 3 in 2-6. Halley interchanged the terms 'greater' & 'lesser' in order to bring these arguments into conformity with the figures. Our solution seems more natural.

5.) There seem to have been schemes of lettering in the figures, particularly in the sequence of figures in the long diorismic arguments. Our lettering has been adjusted somewhat in order to bring this out.

6.) We have introduced some regularity in our translation of the various Arabic words for 'way' in order to clarify what is going on at the conclusion each locus, where Apollonius discusses the locus from the point of view of the variety of ratios.

7.) In the use of connectives, we relied on our English sense rather than literal adherence to the Arabic, because of the fluidity of the particles in any language.

8.) We have deleted the summary argument at the end of the manuscripts because it added nothing new, because the argument pertains to Locus 10 while the figure pertains to Locus 14, and because it is stylistically somewhat different than the rest of the text. It is possible that it is a summary argument for all the most general loci, but this is not clear.

9.) We have introduced the recipient terminology in accordance with the editor's translation of Euclid's *Dedomena* to bring out the logic of the analytical reasoning.

ABOUT THE RECIPIENT TERMINOLOGY

The following remarks are based on the editor's translation of of Euclid's *Dedomena (Recipients, Commonly Called the Data)*.

1. The passive participle of the common Greek verb meaning 'to give' as generally used in geometrical analysis is syntactically different than the passive participle occurring in the enunciations of geometrical problems, to the extent that it is misleading to translate them both by the English word 'given', as is usually done. The situation is particularly confusing, because the two usages may occasionally be mixed together in the same argument, as in the present book.

2. In the statement of a problem, the aorist passive participle occurs in the attributive position, or occasionally in an absolute construction, and modifies the direct object of an act of giving. Since this agrees with English syntax, it is appropriate to translate this participle as 'given'.

3. The primary function of the definite article in these Greek constructions is to specify the attributive position, not to make the noun definite. We feel this when we commonly translate

"On a given straight line_." instead of "On the given straight line_.".

4. The attributive present passive participle occurs solely in the statement of a diorismos. It expresses a state of readiness to be used for some purpose, of 'being offered' for a construction or a synthesis that needs to be performed. A geometrical figure modified by such a participle, such as a straight line, should be thought of not so much as a figure, but rather as marking out a place to be used for the positioning of some other figure, or as the bearer of a magnitude to be used for determining the extension of this other figure.

5. The abovementioned aorist participle is aorist to express entrance into this state.

6. The indirect object of such an act of giving is not explicitly expressed, but is implicitly the figure to be constructed. That is, the straight line is provided to a certain figure, for its positioning.

7. We the constructors are the agents of this act of giving, not some other person posing a problem to us.

8. In this context 'given' has the fundamental sense of the places, magnitudes etc. that we have been able to provide by means of constructions performed on the geometrical figures originally available to us, for the positioning, extension etc. of other figures.

9. In the context of geometrical analysis, the aorist & perfect passive participle occur in predicate position, as circumstantial or supplementary participles. Following the Greek tendency for the indirect object to become the subject of a passive construction, these participles agree with the indirect object of an act of giving.

10. The adverbs 'in position' or 'in magnitude' are attached to these participles. The direct object of this analytical act of

giving is implicit in them. That is, some geometrical figure has been provided for, with respect to its position or extension.

11. The perfect tense suggests that the recipient of an act of giving is in such a state that it is allowed to keep what is already in its possession. That is, its position or extension would not change if it were introduced again in accordance with the same conditions. Thus, this kind of provision is for a preexisting figure, not one to be introduced.

12. When the aorist tense is used in the circumstantial participle, it denotes entrance into that state.

13. The predicate aorist after the finite verb is not so much a supplementary participle as a true predicate adjective, referring to the event that has occurred in the course of the construction.

14. In view of the difference between English & Greek, it seems best to translate this participle as 'recipient', so as not to lose sight of its reference to the indirect object of an act of giving.

15. 'is recipient' translates the perfect participle when it is used as a supplementary participle. 'is a recipient' translates it when it is used as a predicate adjective.

16. To sum up, the two meanings of the passive participle that we have been discussing are not only syntactically different, referring respectively to the direct and indirect objects of an act, but also have to do with fundamentally different acts of giving.

17. Euclid tries to link these two meanings in his first definition in *Recipients*. There he makes the possibility of our providing an equal out of our own resources a criterion for the state of extensional recipience.

GLOSSARY OF TERMS
OF PARTICULAR IMPORTANCE TO THIS BOOK

Analytic Terminology

Analysis translates taḥlîl [√HLL form II, verbal noun] cf. Gk ἀνάλυσις. *Synthesis* is for tarkîb [√RKB form II, verbal noun] Gk σύνθεσις. The same word is used a few times in the expression *by construction*. *Sought* translates maṭlûb [√ṬLB passive participle] Gk ζητούμενον. *Requisite* is for mukâf^in [√KFY form III, active participle] probably Gk ἐπιταχθέν.

Recipient translates both the Arabic word mafrûḍ [√FRḌ passive participle, literally 'assigned'] and ma^clûm [√cLM passive participle, literally 'known']. According to our conjecture, the translator used the first for ἐπὶ δοθέν and the second for ἐπὶ δεδομένον. We translate as *is a recipient* and *is recipient* respectively. A few times the Arabic uses mu^ctà [√cṬW form IV, passive participle, literally 'given']. Here we suppose Gr had δέδοται, which we translate also as *is recipient*. We also use the expression *the ratio being given for the synthesis* for the presumed Greek ὁ διδόμενον λόγος εἰς τὴν σύνθεσιν. mawḍû^c [√WḌ^c] is clearly ἐπὶ θέσει *positioned*, which we sometimes translate as *location*. *Magnitude* in the expression *in magnitude* is qadr [√QDR].

Terms Pertaining to the Division of the Problem and Diorismos.

Locus translates waḍ^c [√WḌ^c] Gk τόπος. It has here its technical analytic meaning, since what distinguishes this level in the division of the problem is the location of the recipient point, namely, which angle it lies within. *Case* is for wuqû^c [√WQ^c verbal noun] Gk πτῶσις. We also employ the word *case* to translate wajh when it refers to cases. The Arabic idiom 'So let the falling of the line in the first way be...' is rendered *So let the line fall in the first way...*

Limitation & *restriction* both translate ḥadd [√HDD] Gk διορισμός. *Limit* is nihâyah. Some of these words may have been ὅρος or even πέρας. *Status* qawâm [√QWM noun], *how* kayfa & *how many* kam may correspond to the πότε καὶ πῶς καὶ ποσαχῶς of Pappus' definition of διορισμός.

The various Arabic words having the general meaning of 'way', wajh, maᶜnà, jihah seem to be used interchangeably. However this causes a great deal of confusion in the concluding section of each locus where the locus is approached from the point of view of the varieties or cases of the ratio. In this instance the word *domain* has been used to make it clear that Apollonius is now talking about these cases.

Special Terms for the Problem of this Book

Cuts off translates qataᶜa [√QTᶜ] Gk ἀποτέμνει. It has been translated merely as *cuts* when the segment is being used to define the particular case under consideration. The *terminal point* muntahᵃⁿ [√NHW form VIII, passive participle] and *the place (point) which is passed by* majâz ' [√JWZ] seem to be synonyms for the fixed endpoint of the segment to be cut off. Cf Gk πρὸς τῷ ἐπ αὐτῆς δεδομένῳ σημείῳ. When Apollonius wishes to say that the drawn line may be anywhere in the case under consideration, he says that it cuts *all along* jamîᶜᵃⁿ [√JMᶜ] a certain line.

General

The manuscripts use two relation words mutasâw[in] & mi_th_l more or less interchangeably. Following Greek usage, we translate them as *equal* to mark comparisons of quantities, and *the same as* to mark comparisons of ratios.

There are two formulaic expressions that occur quite often: 'So let what we have mentioned be according to its state.' and 'Things will be according to their state.' These derive from Greek expressions such as τῶν αυτῶν ὑποκειμένων and τῶν αυτῶν ὄντων. They have been translated with some freedom as the context requires, usually by something like *With the same things supposed,* or *Let things be as they were before.*